Cover design by: www.doodlescreations.com

Dedication

To my amazing wife Tessa and our wonderful
daughters Lucy & Sophie

The future of energy

Praise for 'The Future of Energy'

There are a lot of long, intricate reports, papers and books about the "energy transition" that are rather dull. This is the opposite. Concise while being comprehensive. Thorough but with a bit of a personal perspective that makes it interesting. Realistic about the challenges but with a dose of optimism about what could be done. Well-informed but accessible. However, the only certainty is that this will change over time so I hope the author can offer regular updates. **David Elmes, Professor, Warwick Business School (September 2020)**

I would highly recommend this book to anybody working within energy or interested in learning more about the movement towards clean energy. I had been looking for a book like this for years but could not find anything that was not a chunky textbook. It is very factual, and John also offers his expert opinion on issues. Looking forward to next edition! **Jon (Amazon, August 2020)**

Great book by John Armstrong on the future of the energy transition and I recommend it to anybody who has an interest in this field. Having known John for several years he has a great understanding and excellent knowledge around this sector! **K. Singh (Amazon, September 2020)**

The trouble with a lot of experts is they are not so much helping you learn about a subject; they are using you as a platform to either convince people just how clever they are, or they are pushing that one silver bullet that will solve their favourite soapbox topic. Not here though, the author takes you on a rapid journey through a range of environmental scenarios (home heating, air travel, industrial consumption) where energy is the problem but also, potentially, brings a range of solutions. Enough detail to keep the purists happy but also simply written so the amateur will not get lost. **Andrew (Amazon May 2020)**

Real world experience makes all the difference. I really enjoyed John's book. What sets it apart is the fact that it is written by someone who has worked in the energy industry. What this means in practical terms is that when the author discusses some of the practical issues, he understands the nuts and bolts of the problems and opportunities. There are plenty of books out there on this subject, but few written with the authority of actual real-world experience. Buy this book. **Mark (Amazon, Oct 2020)**

Contents

The future of energy

The future of energy

Acknowledgements

Thank-you to everyone who read the 2020 edition and provided such valuable challenge and encouragement.

1. **Forward**

When the first edition of 'The Future of Energy' was released in May 2020, the world was in turmoil. Countries across the world had locked down their citizens, planes were grounded, city centres lay eerily silenced and roads were empty. In May 2020, the situation of the Covid-19 pandemic was constantly evolving, and it was impossible to tell whether changes would have just a fleeting effect or a sustained impact. A year on, it is now possible to reflect on one of the most profound years in energy and to begin to consider how these and many other changes over the last 12 months will impact energy in the future.

As with most sectors, Covid-19 will leave its shadow for years to come, and one sector most notably impacted will be energy. Whilst some parts of the energy sector, such as oil, were rocked by falling demand and low prices, others remained much more resilient, such as renewables. Previously niche areas of the energy sector had their day whilst traditional high-dividend-paying oil and gas saw record falls in profit.

The pandemic highlights perfectly the challenge of predicting the future. Whilst much forward gazing relies on long-term trends, it is these high impact, low probability events which accelerate change and in some cases trigger tipping points. In January 2020 with Covid-19 a problem happening far away, nobody would have believed that only a few months later cities would be deserted, trains would be running empty, and working from home would be the new norm for many.

Perhaps with hindsight it is now possible to say that a pandemic was not that unforeseeable. The UK Government's own risk register has a pandemic as the highest impact event, and one of the most likely[1]. It is therefore interesting to consider in predicting the future how we must look not just to trends but also to the 'black swan' events that in hindsight seem completely foreseeable but in foresight simply hide in plain sight.

In 2020 I made ten predictions for 2030. I enjoyed plenty of healthy debate on these over two platforms: LinkedIn and Energy Central. The two predictions for energy in 2030 which felt at the time a stretch too far now seem to be increasingly realistic – firstly, 50% of vehicles will be electric, and secondly, there will be a reduction of 50%

in aviation. Reflecting on these, I think the aviation reduction may be less likely. However, with the ban on diesel and petrol cars shifted forward a decade to 2030[2] in just a few months since I made my prediction, my call of 50% electric vehicles by 2030 feels like a safer bet.

We start from an unexpected place in 2021 and in many ways a positive one. The tough challenges of 2020 have given us space to reset, consider and restart in a new way.

In updating this book, I came across the concept of 'green swan', disruptive green technologies hiding in plain sight. I believe that it is likely that the technology which will define the next decade in energy already exists and that in the next two or three years the path ahead of us will become significantly clearer.

2. Introduction

In this book I aim to take the reader through a possible future of energy generation, transportation and utilisation, seeking to make some bold calls on what energy will look like in the next decade and beyond. I do not seek to provide a detailed 'textbook' on the future of energy. Instead, I seek to bring together some discussion on energy and thoughts on a range of topics which for me form the fulcrum of the challenges ahead of us. Although focused on the UK energy system, the ideas are transferable to anywhere in the world.

Energy is a huge field, touching every part of society. Without it we could not cook, heat our homes, make steel, travel or pretty much do anything. Since humans first made fire to warm themselves and to cook, energy has been a cornerstone of progression, and since the times of Watt and Brunel it is hydrocarbons in the form of coal, oil and gas which have driven us forward, forming the cornerstones of a revolution which has changed every aspect of our daily lives.

In 2021 we have started to turn down a new path for energy. We have chosen not to continue our previous path reliant on hydrocarbons and have begun to progress down an alternative route in which we find another way, utilising hydrocarbons differently – and in lower volumes - whilst finding energy from 'alternative' sources including many that already exist and are rapidly moving from niche to mainstream.

There exists a huge range of information on the 'energy transition', with competing technologies and theories vying for supremacy. It is easy to fall into the trap of believing there is a quick answer or 'silver bullet' to the huge challenges we face. It is substantially more complicated with an inevitable patchwork of future technologies, rather than a single simple solution. There is no perfect answer to the challenges we face, but most will in some way shape the way we use energy through the next decade and beyond. One thing we need to remember is that in energy there is no 'free lunch'. No technology is consequence-free, and we must approach choices with an open mind.

There is a raft of excellent information out there on all facets of the energy transition; however, what I do not see is enough open debate about the pros and cons of different options and the true potential of emerging technologies. I believe it is only through open dialogue and discussion that we will find the best path through the complex maze of challenges we must navigate.

Energy is a rapidly evolving and often highly politicised topic, and as such this book is clearly entitled the 2021 edition. Within this book my predictions will inevitably be wrong to some degree, as technologies will advance at different paces and there will be ideas that may change the face of energy forever - ideas which have not even been yet thought of. I do believe that it is in the debate and dialogue about the future that we will innovate and generate ideas together and create amazing outcomes.

I will be back for an update in 2022 when it will be interesting to see whether my predictions are turning out to be broadly right or way off the mark.

This book is designed to question, challenge and raise debate and as such I would welcome your thoughts and comments at the following:

www.johnarmstrong.co.uk/futureofenergy

alternatively, you can email me your thoughts:

futureofenergy@johnarmstrong.co.uk

or contact me on LinkedIn:

www.linkedin.com/in/johnmichaelarmstrong

Scan with the LinkedIn app:

PART 1: SOURCES OF ENERGY

3. Talking carbon and greenhouses

I do not intend to dive into the science behind the climate crisis. The scientific case is compelling, and the impact will shape our planet for generations to come.

I have met several people who doubt the validity of the climate change theory and I have enjoyed several lively debates on the topic. I would, however, say that what matters is not individual opinion but that of society and the governments that run our countries. Opinion has moved substantially over recent years, and further substantial regulatory interventions around greenhouse gas emissions are inevitable. This is hugely important to the world of energy and will have impacts on the way we travel, heat our buildings and run our industries.

To understand the energy transition, you need to get a grip on greenhouse gas emissions and their link to energy. Carbon dioxide is just one of several gasses which cause the greenhouse effect and is responsible for 76% of global warming. Next is methane at 16% and nitrogen oxides at 6%, both of which are emitted mostly during agriculture and industrial processes[3].

When we talk about carbon emissions, some big numbers come into play. Currently the world produces about 33 gigatonnes of carbon dioxide each year[4] (2019 figure). 2020 has been an exceptional year with an estimated 2 gigatonne reduction in emissions[5] caused by global lockdowns and a curtailment of economic growth. Once restrictions ease, the reduction is unlikely to be sustained for long, and emissions will return to 2019 levels and begin climbing again.

This gives a world average of around 4 tonnes of carbon dioxide emitted per person each year. There is a large disparity across the world with, for example, each person in the United States emitting 19.5 tonnes compared to residents in the UK being responsible for a significantly lower 5.65 tonnes. With plenty of other extremes in either direction such as Qatar at 38 tonnes a person and Ethiopia at a tiny 0.1 tonne[6] there are clearly huge differences between each of our footprints on the planet.

I like the current 6-tonne number as it puts into context an individual goal of a net-zero[7] per person target by 2050 (this is the target for most Western nations in order to limit the impact of global warming).

It is impossible to find an average person or an average household, so I thought I would make it personal and share my own. Here are some useful numbers from 2020 for my carbon footprint, which I calculated at carbonfootprint.com. I have compared these to my 2019 figures to see what effect Covid-19 has had on my footprint.

- Heating and electricity: 1.65 tonnes (6 tonnes shared between four people).
  - 15% increase in electricity consumption.
  - 10% increase in gas consumption.
- Flights for business: 0. (0.7 tonnes in 2019)
- Flights for pleasure: one return trip to Belfast 0.5 tonnes.
- Hotels for work: 0.2 tonnes (2 tonnes in 2019)
- Car travel: 3 tonnes – 15,000 miles in a family car. (4 tonnes in 2019)
- Rail travel: 0.2tonnes (1 tonne in 2019)
- Car replacement: I kept running an old car rather than replacing it - around 2 tonnes a year (3 tonnes in 2020)
- Food: 1 tonne
- Other (clothes, insurance, computer equipment etc): 3 tonnes

- Solar panels: negative quarter of a tonne a year (about 1 tonne for the house shared between the four occupants – 5% increase on 2019 caused by the extremely sunny April and May)
- **Total: 11.3 tonnes (16.75 tonnes in 2019)**

Despite increased home-utility use of about 10%, I have seen an overall 30% reduction in my personal carbon footprint between 2019 and 2020 - driven by travelling a lot less and hanging on to a car rather than replacing it. I am still, however, smashing through the UK average of 6 tonnes, and my halo from thinking by being a Brit I was not as bad as the gas-guzzling Americans remains substantially dinted.

These numbers are useful when you consider the potential for a personal carbon budget of zero tonnes by 2050. Even if I could sustain my 2020 reductions, I am going to have to make some big changes or I am going to have some forced on me.

72% of greenhouse gas emissions come from energy production.[8] If you look at my personal numbers above, it is clear just how much comes from transport and energy for the home. These two areas alone make up 60% of emissions for my life (and I suspect are feeding into the emissions of the other areas as well, such as hotel stays, etc.). Put simply, if you do not crack energy, you do not crack climate change.

Having a good conversation about carbon matters and completing a personal carbon assessment is a great way to get a handle on your own impact. If you cannot measure it, you cannot control it.

Have a go at your own annual carbon review. I will come back and track mine in the 2022 edition.

4. Going green – renewable sources of energy

There is no such thing as guilt-free anything – however, some sources of energy are greener than others.

When we discuss renewables, it is important to be clear that these are forms of energy through which no carbon is emitted by their production and utilisation. Examples of renewables are wind, solar and biomass – although biomass is a little more complicated and the subject of a later chapter. By 2035 it is projected that globally 50% of energy consumed will be from renewable sources[9].

What is the potential for wind energy?

Between 2010 and 2020 there has been a staggering increase in global wind capacity, with 650 gigawatts of global installed capacity in 2020, growing by around 60 gigawatts a year[10]. To put that size into context, a typical large coal-powered station is around 2 gigawatts.

In the UK, the growth of offshore wind and its cost-competitiveness against fossil fuels is credited with spurring a substantial decline in carbon emissions for electricity generation over the past decade (making most of the contribution to the UK's overall carbon reduction achievements). In 2019 20% of UK electricity was generated by wind[11], and as we moved in to 2020 the first quarter number was a staggering 30%.[12]

The current trend in offshore wind is for turbines to get even bigger, and with more areas of the seabed being offered up for development all the time, the stage is set for super-growth in the sector. Globally, wind capacity is expected to go up by an incredible 112% over the next decade[13], growing from 29GW capacity now to well over 234GW in 2030[14]. The UK government recently has stated a goal to quadruple current offshore wind capacity of 10GW to a staggering 40GW by 2030.[15]

Wind energy does, however, come with some challenges.

- It is not always windy. There are some days where it is not that windy at all. On 5th January 2019 wind assets managed just 142MW of generation. Compare that to the windiest day in 2019, 10th December, when at peak a staggering 13 gigawatts was being produced in the UK[16]. At scale, however - for example, across a whole nation - wind is reasonably predictable, so with the right kit it is possible to be able to plan for supply changes as the wind comes and goes, making use of other fuels as back-up or for storing energy for future use.

- The wind is not always where you would want it to be. The least windy town in the UK is St. Albans to the north of London. The windiest place is the Shetland Islands. With most of the 'good' wind available offshore in the north and west, that energy needs getting to the large cities, most of which are inland or in the south. Wind will inevitably form part of the drive to decarbonise electricity globally. It is, however, not the only answer, as its intermittency and geographical challenges do not quite make it the 'silver bullet' for our energy problems.

The above two challenges, however, are offset by ongoing cost reductions for offshore wind, making the technology compete head-to-head with hydrocarbons[17]. With its completely green energy and an increasingly improving economic case, offshore wind is set for huge growth over the next decade.

Finally, for completeness it is important to reference the carbon produced in manufacturing and installing the turbines. Various studies have put the carbon payback time of wind turbines at around 9 months[18] with a lifecycle carbon cost of 12 gCO2eq/kWh[19], which compares favourably to coal or gas powered stations of 786 kG CO2eq/kWh and 365 kCO2eq/kWh respectively[20]. Do not get too worried about the units on these; these figures are just useful to compare the carbon intensity of different technologies.

Use the sun

Solar energy has the potential to provide an abundance of energy, and with 623GW of total installed capacity, it is the most installed renewable technology.

Demand - much of it driven in the early stages by government subsidies - has prompted growing interest and economies of scale, making solar photovoltaics accessible to millions of homes as well as industrial scale solar farms. The total installed UK capacity is around 13 gigawatts.[21] That is the same as 6 or 7 large coal power stations at full power.

However as green as solar is, it does come with some inherent challenges in operating the system.

- The sun does not shine for a lot of the day. In fact, solar very rarely runs at its peak potential in the UK (which is not to say it does not work – photovoltaic panels run on daylight rather than direct sunshine so there is always something being produced during daylight hours).

- Solar panels take up a lot of space. Unlike wind, which can be parked out of the way offshore, solar panels take up an awful lot of land and have specific geographical needs to run at their best. There are simply not enough (south-facing) roofs for all the solar panels needed. Added to that, use of land becomes a topic of strong debate as our open spaces come under increasing pressure as populations grow.

Like offshore wind, and despite the challenges above, solar has huge potential. Once again costs have plummeted such that the technology can compete head-to-head with hydrocarbons. A huge milestone achieved recently is the first projects emerging which do not rely on any subsidy from the government,[22] a clear sign that solar is set for fantastic growth and is a technology to stay.

Are there any other sources of truly renewable energy?

Other interesting opportunities for completely renewable energy are out there:

Holding back the tide

The idea of generating electricity from capturing tidal movements remains tempting. These projects have the potential to deliver around 20% of UK energy demand (government numbers)[23]. However, the environmental impact of building such a project remains significant. With several projects in the pipeline, I believe that one of them must get traction and take off; the lure of predictable zero carbon energy forever must surely become too much to do nothing.

Going surfing

Early excitement about energy from waves has dwindled as harsh marine conditions and poor economics have seen off early contenders.

There is, however, some hope of development of this technology, as it has the potential to deliver consistent base-load energy supply, unlike the current major renewable contributors, wind and sun.

Drilling deep under ground

Geothermal energy has some huge potential in locations with easy-to-access heat underground. The Eden Project in Cornwall with its 5km deep bore-holes is a fantastic demonstrator[24]. If costs of bore-holes can be reduced, then geothermal has the potential to be a game changer for us all. The economics, however, do not currently work for widespread use.

Biomass & biogas

Solid biofuels already make up around 8 percent of energy production globally. That is a lot of wood. In many developing nations wood is the primary energy source for heating and cooking. In developed nations wood historically has been relegated from homes but has recently seen a resurgence in power generation. The potential for biomass and biogas is covered in more detail in a later chapter.

5. Do fossil fuels have a future?

Fossil fuels present both a challenge and an opportunity in decarbonisation; they provide exceptionally good vehicles for moving and deploying energy, not least due to the substantial amount of existing transport infrastructure in place but also due to their extremely high energy densities (how much energy is in each kilogram).

With increasing focus on renewable technologies, you would think the days are numbered for fossil fuels. However, that would be an easy mistake to make, as even in the most ambitious projections for reducing carbon, fossil fuel-based sources of energy still account for 77% of world energy production in 2040 – it is therefore quite clear fossil fuels are with us for quite a while.[25] In fact, exploration of oil continues at a fast pace, with new reserves being identified and developed constantly.

Hydrogen

In the chapter on hydrogen, I discuss the role of hydrogen in a future energy system. Blue hydrogen, where hydrogen is created from methane and the carbon dioxide that is produced is stored underground, provides a clear pathway for hydrocarbon use to continue whilst minimising environmental damage. Furthermore, if blue hydrogen is made readily available, then there is the option to reform this hydrogen into 'synthetic' liquid fuels.

I will discuss the role of hydrogen and synthetic fuels later in the book.

Carbon capture and storage

Perhaps the most enticing technology for decarbonisation is the idea of carbon capture and storage. The idea is to capture carbon dioxide at the point of release (burning in power stations or industrial processes, for example) and instead of sending it into the atmosphere, pump it deep underground and store it under pressure. This is a proven technology and could enable even the dirtiest of fossil fuels (i.e., coal) to have a future. However, the costs of the process have to date made carbon capture and storage an enticing and yet unfulfilled technology.

Processes that need a lot of heat

Some things just need a lot of energy to happen. Steel and glassmaking, for example, along with a range of other industrial processes, require high temperature, high intensity heat. There is an argument that hydrocarbons will still need to be used to support these types of processes (and potentially rationed to ensure these processes can continue well past 2050). It is in this area that some kind of localised carbon capture and storage could be needed.

Applications where nothing else will do.

Oil and gas provide the feed stocks for a huge number of processes where no alternative currently exists. Materials such as plastics need hydrocarbons as a feed stock, although alternatives are in development.

Some interesting examples of products currently dependant on oil are[26]:
- Plastics
- Tyres
- Antiseptics
- Deodorant
- Footballs
- Soap
- Refrigerant
- Insect repellent
- Candles
- Anaesthetics

- Toothpaste
- Heart valves
- Glasses
- Ink
- Asphalt (65 Kg C02/Km of road[27])
- Lots more…..

There are some interesting products on this list like anaesthetics and antiseptics that we would certainly need to replace with a suitable lower carbon alternative! As an aside, anaesthetics have a higher greenhouse-causing effect than carbon dioxide. These gases are far more damaging in their use than in their manufacture[28]. An hour of surgery releases 24Kg equivalent of CO2[29]

Finally, an interesting challenge of reducing natural gas production would be a shortage of helium – a by-product of the natural gas extraction process. Perhaps mildly inconvenient for our party balloons but quite a major issue for Magnetic Resonance Imaging (MRI) machines in hospitals and similar applications, which rely on the element to super cool magnets[30].

The future of energy

6. Lessons from the golden age of steam

Energy assets have long life expectancies - really long. Decisions made today to build assets impact the energy system decades in the future. Looking back to early in the last century, a huge transition happened in rail; the move from coal-powered steam trains over to diesel and electric took the best part of 60 years, with dated technology being constructed long after its demise was certain. It was obvious from the late 1940s that steam's days were numbered, and yet it was not until 1960 that the last steam train left Swindon's incredible 300-acre[31] factory, with over 200 built during the preceding decade. That train was still running commercially up until the day steam trains were banned in 1968.

Reflecting on this transition, it is interesting to look at the dates and consider some of the challenges we now face in transitioning energy across the globe.

- 1814 – First commercial steam train[32]
- 1879 – First electric train (built by Werner von Siemens)[33]
- 1925 - First commercial diesel train[34]

- 1930s – First diesel trains operating in the UK.
- 1960 – Last steam train built in Swindon *(The Evening Star)*[35] with 200 being built in the decade before.
- 1968 – Last commercial steam train taken out of service.[36]

Between the first commercial diesel train and the last steam train being retired took 43 years. It also took legislation to force the final steam off the rails rather than the asset reaching the end of its useable life. It is interesting to think of all the infrastructure required to keep steam trains on the tracks - coal provision, watering (an impressive 22,000 litres every 100 miles!)[37], along with all the supplementary maintenance required for the technology. For steam trains it took a change in law to get the last one off the rails - not the appearance of a superior technology.

What is different?

A lot is different now! For a start there are many more people on the planet (7.8 Bn vs 2.3 Bn), so the impact of the use of different types of technology is far greater. Most importantly, the speed (and volume) with which we can now communicate is disproportionally faster and greater. Those were the days of mail trains and telegrams - not Whatsapp and Tiktok! Digitalisation simply enables technological change to happen much faster.

What is interesting however is how digital change outpaces the fundamental life of equipment - energy infrastructure has lifetimes that far exceed the obsolescence of digital systems. A small but local example for me is the awful sat nav in my four-year-old car – in such a short time the technology provided in the vehicle four years ago is now clunky and unusable whilst the vehicle itself keeps on running.

Operational life of new assets today:

- Domestic Car: Around 12 years[38] (electric car batteries last about 10 years)
- Domestic Gas Boiler: 12-20 years[39]
- Offshore Wind: 25 years+[40]
- Solar Panels: 25-30 years[41]
- Oil Rigs: 40+ years (According to the *Guinness Book of World Records* the oldest is 70 years!)[42]
- Nuclear Power Station: 50-70 years[43]
- Coal Power Station: 50+[44]
- Gas and Electricity Networks: 50+ years

What does this mean for the energy transition?

The basic engineering of the ageing of metal has not changed a great deal – if anything we now have the technology to make things last. That means decisions need to be made in the context of long-time horizons; the last steam train did not come off the rails because it got too old – it came off because it was pushed. None of those 200 steam trains to leave the Swindon works between 1950 and 1960 delivered their economic life, so why did they get built? Answering that question helps us understand how we transition more smoothly and more quickly this time.

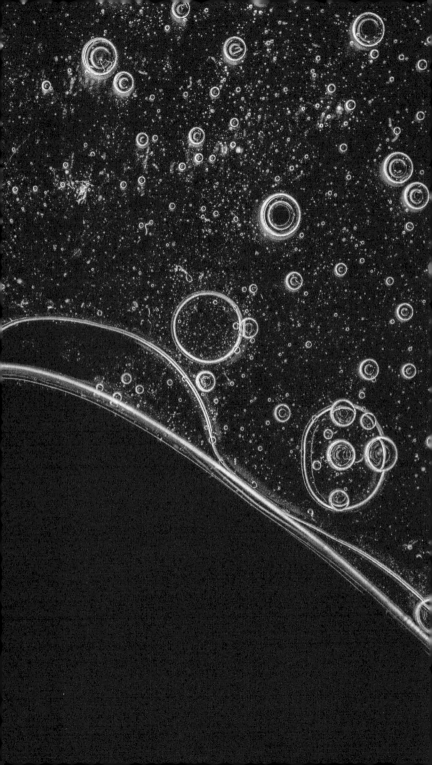

7. Is the future of fuels synthetic?

I have highlighted the challenges which batteries and hydrogen have with energy density and even suggested ammonia as a possible alternative. There is one tantalising prospect which I have not yet addressed, and that is of synthetic fuels.

The premise of synthetic fuels is that a liquid hydrocarbon can be formed from either biomass or the reaction of hydrogen with carbon dioxide. Currently, alternative liquid fuels are generally biomass derivatives. It is, however, the fuels made from pure hydrogen, called e-fuels, which provide an exciting prospect for decarbonising sectors that need a high-density source of energy.

The 2019 BP energy outlook puts global annual demand for transport fuels at just under 2000 million tonnes in 2035[45] - a gradual climb from the 1500 million tonnes consumed now. The lion's share is taken up by trucks and around a third by aviation and marine transport.

Trucks, marine and aviation present the biggest challenges to decarbonisation – currently there is no scalable green alternative primarily due to the energy density of the alternate sources of energy available. It is this energy-density challenge that makes e-fuels present a potentially exciting solution.

How are e-fuels made?

To make e-fuels you need plenty of pure hydrogen and carbon monoxide. Carbon monoxide is made by passing carbon dioxide through the excitingly named 'reverse water gas shift' reaction, which uses electricity to split up the carbon dioxide molecules[46]. There are plenty of sources of carbon dioxide out there, such as burning methane, coal or even biomass. The benefit of using carbon dioxide produced during combustion in this process is that it would not be released into the atmosphere and could be used again.

On the other hand, hydrogen, required in large volumes, would be more costly to find. However, in a hydrogen-driven world the prospect of plentiful 'Green' hydrogen produced from offshore wind through electrolysis is the most hopeful source. Alternatively, both the hydrogen and the carbon dioxide could be obtained by steam-blasting methane. Wherever the hydrogen comes from, the process is likely to be very energy intensive!

The final e-fuels process is called the Fischer-Tropsch[47] process. This process needs a catalyst such as cobalt, ruthenium or iron. The feedstock of pure hydrogen and carbon monoxide reacts with the catalyst in a chamber under pressure – the output being pure liquid hydrocarbon. This process generates a lot of heat, and it is this heat generation that makes the process quite inefficient, as much of the energy available in the hydrogen molecules is lost as heat. If this heat can be used, for example in an industrial process or district heating, then there is the opportunity to improve overall system efficiency.

Will e-fuels feature in the future?

Studies such as that by the Royal Society have put costs at around 1.5 euros a litre by 2050[48]. However, the current price tag of 4.5 euros a litre prices them out of the market.

If legislation pushes trucks, aviation and marine to zero carbon, then I believe that synthetic e-fuels will likely form part of the story. To what degree will depend on the depth of decarbonisation and how competing technologies can push the energy density frontier. Additionally, human behaviour will impact demand for applications needing high energy densities: Could road transport be shifted to electrified rail? Will there be less long-haul travelling?

Finally, it is worth noting that catalysts used in the process are hard to find and come with their own challenges. In particular, cobalt mining has some important ethical considerations as does nickel, of which ruthenium is a by-product. It seems that as with battery technology, all pathways need considering, to be both low carbon *and* sustainable.

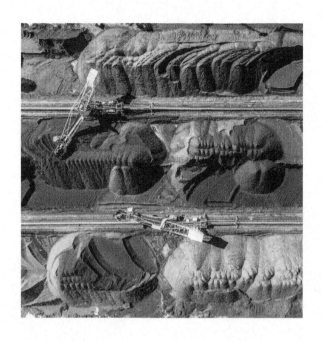

8. Fission, fusion or modular – the future for nuclear

Nuclear energy is one of the most intriguing future energy technologies, most notably because it simultaneously has the most potential to meet our low-carbon energy need, but also because it may just be the scariest and most divisive (although check out the chapter on ammonia).

There are two extremely different developments in nuclear, both of which have the potential to revolutionise the way we live - one 'centralised' and the other very much 'decentralised'. They are, however, two technologies that have been 20 years away for the last 20 years - not least as the public's perception of nuclear has become increasingly negative over that time.

The total installed capacity of nuclear power stations globally is 460 GW (that's about 10 percent of global power demand), capable of generating nearly zero carbon electricity (there will, of course, be some carbon emission from mining the core fuel, as well as the staggering amounts of concrete used to build nuclear power stations). These 450 nuclear power stations use nuclear fission (splitting the nucleus of an atom)[49], the downside of which is quantities of hazardous waste, which is tricky to manage, along with processes which need to be carefully controlled to avoid exponential reactions, which can cause widespread damage.

Nuclear fusion, however, is interesting. With fusion we are trying to replicate the reaction which happens in the sun. Put simply if you take two hydrogen nuclei and throw them together you get helium. Unfortunately, it is just not that simple. There is, however, a huge amount of research into fusion, and it has the potential to create guilt-free energy (no waste and no risk). There is one small problem and that is that no one has figured out how to do it on a commercial scale yet. That is probably 20 years away.

The next nuclear technology which is interesting is small modular nuclear reactors (SMRs). Currently, most nuclear power stations have one large reactor and generate a couple of GW of electricity. To house all the safety equipment and various processes around nuclear generation, you need economies of scale to make it all work.

That, however, is not the end of the story. For decades small reactors have been used on the high seas - most notably in submarines, aircraft carriers and, more recently, ice breakers designed to break through thick Arctic ice.

These small reactors present an interesting opportunity for zero carbon energy, literally on our doorsteps. Small reactors in development are typically sized around three hundred megawatts as opposed to the current larger scale reactors at sixteen hundred megawatts. These present an enticing opportunity for more decentralised power production using technology which already exists but has not been deployed in such a non-military or on shore application.

There are of course significant challenges away from technology development as the public perception of nuclear energy – never overwhelmingly optimistic at the best of times – has waned further following incidents such as Fukushima. It is an understatement to say the idea of a nuclear reactor on your doorstep may be a difficult sell to host communities.

Despite the above obstacles the UK government has committed in 2020 to develop the technology and has included it in its energy strategy. It will be interesting to see if public sentiment can be changed and got behind this exciting technology.[50]

9. Biomass and biogas

Solid biofuels - for example, wood - already make up around 10% of energy production globally[51]. That is a lot of wood. In many developing nations wood is the primary energy source for heating and cooking. In developed nations wood historically has been relegated from homes but has recently had a resurgence in large-scale power generation.

Biomass (chopped up trees or waste wood)

Biomass presents some enticing yet challenging opportunities for alternative sources of energy. I have long struggled with the idea of chopping down beautiful trees, chipping them, shipping them across the world and then burning them for electricity – although I can see the logic. I probably just like trees too much. The logic of biomass generation is that the carbon dioxide which the tree absorbed as it grew would be released as the tree rotted, so by releasing that carbon dioxide by burning it, we are effectively skipping a step and releasing the carbon dioxide through combustion rather than rotting - the key, of course, being that you re-plant with a new tree afterwards.

In the UK biomass has been promoted heavily through the 'Renewable Heat Incentive' (RHI) subsidy and some large biomass plants have been built.

Biogas (from rotted waste)

Biogas is a hugely different prospect. Typically, biogas is made by rotting some form of waste (waste food, sewage, cow poo).

Recently, the largest biogas market has been through capturing the gas from existing land fill sites and then burning it locally to generate electricity. This absolutely makes sense, as the methane emitted from the land fill site has a far higher propensity to cause global warming than the carbon dioxide emitted once it is burnt, and the production of electricity at the same time makes it a double win. Biogas from household refuse or farm slurry has some fantastic potential. It does, however, take a lot of treatment to get it to a state to be injected into the main gas grid.

How is biogas made?

To make biogas the waste material is biodegraded in an anaerobic (without oxygen) atmosphere. Done right, this emits plenty of methane. This is basically what is happening in the big green tanks you might see at the side of the road as you are driving around the country.

Once the material is generating plenty of methane, the gas needs to be treated and mixed with some other gases to enable it to be fed into the main gas grid. An alternative is to burn it locally for local heating or power generation. This saves the costs of transporting as well as treating the gas for mainstream consumption.

Are there enough cow pats to go around?

With biogases the main limiting factor is not from the technology but from the volume of suitable waste available to process into gas. Future projections for the UK all show biogases having a role to play in provision of heat – but not in a significant percentage.

What about trees?

Biomass with an appropriate supply chain has a far higher potential than biogas. However, it is essential that any biomass supply chain is robust and able to deliver carbon benefit whilst not adversely impacting local ecosystems where the trees are grown.

Chopping down trees to save carbon remains a difficult choice and one which is likely to remain controversial.

10. Carbon capture and storage

The potential to capture the carbon released from combustion and store it in some form is enticing. The idea and technology have been around for some time, with carbon dioxide used to recover more oil from as far back as the 1970s.

Catching the carbon dioxide before you burn the fuel (pre-combustion)

Several technologies exist to capture the carbon dioxide in a chemical process, a good example being in the manufacturing of hydrogen from methane, where the hydrogen molecules are split away from the carbon. Once the carbon is removed, the hydrogen can be combusted without environmental harm.

Catching the carbon dioxide after burning the fuel (post-combustion)

A much-discussed technology for older coal fired power stations was that of capturing the CO2 after combustion. This means keeping the existing technology and installing some 'cleaning' technology on the bang end. My favourite description of this is that it is like having an old banger with a shiny exhaust! In this case a chemical solvent (likely amine) is used to absorb the CO2 from the flue gas[52]. By warming and cooling the solvent [53]the CO2 is released in relatively pure gaseous form. It can then be piped to a suitable store.

Transporting carbon dioxide safely

Carbon dioxide can be compressed to near liquid form and pumped safely over long distances. If emissions sources could be linked together, then networks could be used to effectively 'reverse' the existing hydrocarbon distribution processes. Existing oil and gas infrastructure could be reversed (with plenty of engineering to make it work!) to enable hydrocarbons to become circular with emissions being returned to the beginning of the process.

Safely tucked away deep underground

Porous rock deep underground, potentially in oil reservoirs, could be used to store compressed carbon dioxide gas. Studies have suggested that billions of tonnes of carbon dioxide could be stored in this way[54]. Various studies have been completed to confirm whether the carbon dioxide could be hidden away forever or if all the hard work would be wasted. The technology has the potential to ensure that once compressed and pumped deep underground, the carbon dioxide would remain there for 10,000+ years.

Are negative emissions possible?

Optimistically, this technology could be used to achieve negative emissions. Through capturing carbon dioxide directly from the air or through burning biomass (trees that have absorbed carbon dioxide as they grow), a carbon-negative process could be achieved. Not only could this technology be used to limit emissions from our existing energy use but also from our past. In this case the technology very much exists. What remains is to establish an economic case.

The future

The UK government has set out a plan to remove 10 million tonnes of carbon dioxide through carbon capture and storage by 2030[55]. This number sounds impressive until you put it next to the 3400 million tonnes emitted by the UK between now and then, and it's just 0.29% of total emissions. For carbon capture and storage to make an impact in overall emissions, goals are going to have to be much more ambitious.

The future of energy

PART 2: MOVING ENERGY AROUND

11. Storing energy - beware hungry cannibals!

In the chapter on energy generation, I touched on some of the challenges of lower-carbon technologies. Unfortunately, most low-carbon technologies generate their energy at the wrong place or at the wrong time.

That is where storage comes in, and several opportunities exist to store energy for use when it is needed. These range from multi-megawatt, large-scale pumped-water hydro schemes to significantly smaller, single-digit kilowatt domestic batteries. All seek to address the challenge that energy is often needed at times of day when generation sources like wind, solar, or tidal may not be able to produce it.

Storage is a broad topic with requirements ranging from 'inter-seasonal', in which energy generated in summer - for example, by solar - is stored for the darker winter months, to 'intra-day', where energy is stored in the daytime for use in the evening, to finally the 'reactive', where energy is stored sometimes only for seconds to manage tiny peaks and troughs on the system and ensure that system 'frequency' is maintained. The solutions used for storage depend very much on the application required, with some technologies lending themselves to specific challenges.

Pumped water

One of the oldest forms of energy storage is pumped water storage. Put simply, when electricity prices are low, water is pumped up a mountain to be later released down through a generator when prices are high. An example is Dinorwig in North Wales[56] with its sixteen kilometres of underground tunnels capable of generating an impressive 288MW of power when in transmission mode. Impressively, the Bath Country-pumped storage facility in Virginia has a staggering capacity of 3 gigawatts and can boast that it is the largest 'battery' in the world.[57]

These types of projects have huge potential; however, where open 'lakes' are used they may cause substantial environmental damage if built now. Dinorwig managed to get around the environmental impact by hollowing out huge caverns in a mountain. This kind of project, however, comes at substantial cost and there are no new pumped hydro projects in the pipeline in the UK.

Large, pumped storage projects lend themselves to longer-term storage requirements across days, weeks and possibly even seasons.

Large 'centralised' batteries

Recently large batteries have started to appear across the UK electricity network. There are presently around ten gigawatts of applications in the pipeline.[58] That is a staggering shift from a standing start of zero only a couple of years ago. The predominant technology is lithium ion, with these containerised units set to pop up all over the country. There are, however, several other battery technologies ranging from traditional 'lead-acid' like in a car to more expensive 'solid-state' technologies used in mobile phones and electric vehicles.

Currently in the UK the economics of larger batteries are such that the sector is seeing phenomenal growth.

Small 'decentralised' batteries

In home energy, storage has potential when linked to a domestic solar panel. Some predictions suggest that across Europe installations could rise to around 500 megawatts of installed capacity a year[59] by 2024.

Domestic batteries have the potential to optimise income from solar panels for owners whilst limiting peak demand and supporting the local electricity network.

If there may be negative power prices on the electricity system as in April 2020[60] during the corona virus lockdown, then batteries also give the owner the potential to be paid to import when prices are negative and then export when the power can be sold for profit at the right times.

Domestic batteries, however, remain costly. When I looked for my own system, the payback on a 6 kilowatt-hour battery linked to my solar panels was around 10 years, which felt like a big gamble on whether I would stay in the same property to get payback.

One exciting area which has the potential to be quite disruptive is the use of the batteries in electric vehicles to provide decentralised storage. Potentially, owners may prefer to keep vehicles 'ready to go' rather than allow power to be used for the grid; however, if the price is right, they may go for it. This technology may more likely be driven by the large fleet owners, who have hundreds of electric vehicles charging overnight.

Chemical reactions

There are several chemical reactions which can be used for energy storage. Linked to the chapter on hydrogen, one option is to use excess power to convert water into oxygen and hydrogen. These two gases can then be stored, in liquid form, converted to ammonia or stored under high pressure ready to be either combusted or put through a fuel cell when power is needed.

Any chemical reaction which can be driven by electricity (i.e., through electrolysis) has the potential to store energy, and these reactions are the subject of a great amount of research.

Other ways to store energy

There are many alternative ways of storing energy under consideration such as flywheels, liquified air, compressed air, and capacitors, amongst others. All have the potential to fit into the storage equation somewhere.

Beware of hungry cannibals.

Batteries are an interesting technology. In a free market any form of storage technology works on the premise that prices vary and that the storage can respond by buying when prices are low and then selling when prices are high. This is regardless of whether it is a 1-second 'balancing' transaction, or 100 megawatts of energy stored in summer to sell in winter. When prices fluctuate, storage can make money. The enemy of profitability for storage is flat markets where demand and supply are met equally.

Storage is an example of a cannibalising technology. The first storage to appear in the system when prices are fluctuating wildly makes the most profit. As more and more stored energy joins the system, each new entrant means that all participants make a little less, until ultimately no one is making any money. In this scenario the system is, however, perfectly balanced.

The future of energy

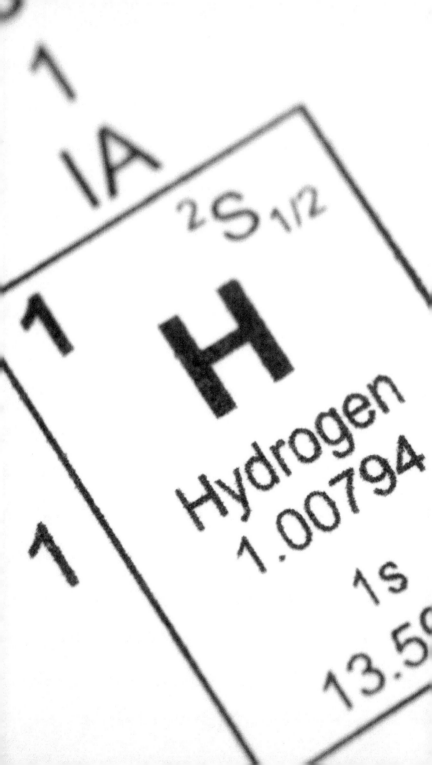

12. A hydrogen future?

The potential for our future heating needs to be met by hydrogen is increasingly widely discussed. Others, however, suggest that increased electrification and improved efficiency mean that a complex hydrogen switchover may not be needed.

The one thing all agree on is that something must be done to move from the current reliance on natural gas for heating. The recent ban on sales of gas boilers in the UK from 2025 is a huge step[61] – however what will replace them remains unclear. Also, the ban does not address the existing housing stock which remains stubbornly reliant on natural gas.

Proponents of hydrogen believe it is the ultimate route to zero-carbon heating, with trials and studies already in full swing. Others such as the Committee for Climate Change see hydrogen playing a potentially smaller role in the future energy system – focused only on niche sections of energy provision[62].

However, even if you believe it is technically possible, there remain some big challenges to a hydrogen switchover. Whilst engineers are focused on solving the technical problems, it is also necessary to explore the social, economic and environmental challenges. Consider, for example, the following questions:

- Who pays for a new boiler, cooker, gas fire (existing ones simply will not work) in every property? What if a resident wants a nicer-looking cooker or fire than what is on offer?
- How do we manage the switchover to a new fuel for large swathes of customers simultaneously whilst looking after the most vulnerable and limiting disruption?
- How do we deal with regional pricing (will hydrogen users pay more or even less per unit of heat than natural gas customers or heat pump users?)
- How will local air quality be managed? (hydrogen burns hotter, so it will produce even more harmful nitrogen oxides than gas boilers do now).

I absolutely believe engineers will be able to produce a reasonably priced domestic hydrogen boiler and that many of the safety challenges can be overcome. However, I suspect the factors outlined above, and others I have not spotted, will make delivering a move to hydrogen at scale far harder than is currently foreseen.

I have seen many an article proudly declaring that this latest hydrogen powered boat/car/lorry is green. We need to be careful with how we talk about hydrogen as a 'green' fuel. Hydrogen is only a carrier of energy and is not in itself green. Seven 'flavours' are used to describe how hydrogen is made and therefore how green it really is.

These are described as[63]:

- **Black or grey hydrogen:** made from natural gas or coal usually in a process called steam reformation. (For the chemists out there, take methane, throw some steam at it and you get carbon and hydrogen). This is super carbon-intensive. I recently read that to decarbonise the current global production of hydrogen used in industrial processes, it would require the entire renewable electricity generation of the European Union.[64]
- **Blue hydrogen:** Pretty much black hydrogen but you find a way to store the carbon dioxide deep underground through carbon capture and storage.
- **Brown hydrogen:** black hydrogen but using even more carbon intensive lignite. This can be turned blue with carbon capture and storage.

- **Green hydrogen:** made through electrolysis using zero-carbon electricity (from nuclear / wind / solar) to split water into hydrogen and oxygen.
- **Pink hydrogen:** like green hydrogen but with the electricity coming from nuclear power stations.
- **Yellow hydrogen:** like green hydrogen but with all the electricity coming from solar.
- **Turquoise hydrogen:** made through methane pyrolysis. Instead of emitting carbon dioxide as a gas, the by-product is a solid. This technology is still under development.

Hydrogen presents both an opportunity and a risk. Existing projects to mix hydrogen with the existing gas network present an opportunity to explore the potential of a hydrogen economy, but without a clear path to delivering green and blue hydrogen, these projects risk exacerbating an existing problem.

In Autumn 2020 the UK government announced a plan to have 5GW of hydrogen production by 2030[65]. This will require substantial investment, although when put next to the instantaneous energy-demand of peak heat (more than 300GW), it feels slightly unambitious.

13. Some like it hot: Is hydrogen the answer to those needing it a little warmer?

An often-overlooked group in our drive towards zero-carbon heat are industrial processes that need higher temperatures. Examples are steel-making (over 1000°C), glass (melting temperature over 1400°C) and even plastics recycling. Heat pumps and heat networks only get you to sub-100 degree temperatures and certainly won't cut it for super-high temperature industrial processes.

Substitution of products is an option – using less steel or more sustainable building materials, for example. However, with increasing urbanisation the reality of the global economy is that there will remain a high demand for those materials used in building the cities of the future (steel, concrete, glass, etc).

Can hydrogen get hot enough?

This is where there is real potential for hydrogen as a fuel. Hydrogen burns up to a cosy 2800° Celsius (about 700° Celsius simply burned in air), giving plenty of opportunities for supporting those needing something a little hotter. It can also be compressed and stored so has potential for transport, particularly in freight.

We may be better off focusing our efforts on processes where hydrogen's potential can be realised – rather than on areas where other technologies are already proven (i.e., domestic heating where heat pumps and heat networks already show us an achievable pathway).

Couldn't we use the sun?

An interesting technology backed by Bill Gates[66] is a concentrated solar system from a company called Heliogen[67]. The system works by concentrating sunlight onto a single point to achieve temperatures in excess of 1500 degrees Celsius. The company suggests applications such as manufacturing concrete, steel and petrochemicals, and treating waste. This technology has recently been named a best invention of 2020 by *Time* magazine, certainly making it one to watch[68].

The future of energy

14. Could 'dirty' ammonia be the clean fuel of the future?

Globally, the ammonia market is worth about 33 billion dollars, accounting for a staggering 1.8% of global carbon emissions[69]. Produced under super-high pressure from hydrogen and nitrogen, ammonia production is hugely energy intensive and more than a little bit 'dirty' when it comes to carbon emissions. Ammonia in liquid form provides a tantalising potential for a green fuel, as it has a relatively high energy density of 3.4kW/litre, even if it does have a few challenges!

Could ammonia really be the fuel of the future? One of the biggest challenges of hydrogen as a fuel is its density in gaseous form along with the high costs of storing and transporting it. Ammonia in liquid form provides a tantalising potential solution, even if it does have a few challenges (being explosive, toxic and very smelly). Potential early applications for ammonia as a fuel include shipping and maybe even aviation.

I'm starting to think that maybe ammonia has the potential to be a 'black swan' technology in some specific sectors.

A 'black swan' event is something which is highly consequential but unlikely. These are often easily explainable – but only in retrospect.[70]

How is ammonia made?

The feed stock for ammonia is simply nitrogen and hydrogen. Nitrogen can be relatively efficiently extracted from the air through cryogenic processes (cooling the air so nitrogen drops out), whereas hydrogen is a bit trickier, as I mentioned in previous chapters.

To create ammonia, nitrogen and hydrogen are fed into a reactor and compressed to a high pressure in a process which has changed little in 100 years, called the Haber-Bosch process. In this process the hydrogen atoms from the feed stock are combined with nitrogen from the air. All of this is quite energy intensive, as energy is needed to extract nitrogen as well as pump up the two gasses to such high pressures.

What about safety and the local environment?

Ammonia does come with its challenges. A quick web search on risks around ammonia gives you this:

'Exposure to high concentrations of ammonia in air causes immediate burning of the eyes, nose, throat and respiratory tract and can result in blindness, lung damage or death. Inhalation of lower concentrations can cause coughing, and nose and throat irritation'[71]

So not the nicest chemical to get exposed to. On top of the potential health impacts it is also corrosive and explosive, so it presents some interesting challenges to engineers.

However, engineers are well used to handling such challenging chemicals. For example, in the United States there are already 2000 miles of ammonia pipelines along with a substantial infrastructure for ammonia manufacture and distribution. So it is certainly not impossible that ammonia might be the right 'green' fuel for some applications.

In fact, existing hydrocarbons are distributed via pipelines, trucks, petrol stations, etc. Petrol and diesel come with their own challenges in terms of environmental and explosion hazards.

Finally, ammonia is regularly used in refrigeration and heat pumps so it is highly likely we are passing near safely stored ammonia regularly in our everyday lives at sites such as hospitals and supermarkets, amongst many others.

This of course all depends on if the feedstock of hydrogen and nitrogen can be generated with no carbon emissions. Without a pathway to zero-carbon hydrogen and nitrogen, there is no zero-carbon ammonia.

Would ammonia be more expensive than hydrogen?

A study by the Royal Society[72] highlighted substantial savings using ammonia in rail, shipping and heavy road transport - in fact, nearly 80 percent less than that of hydrogen gas and a third of the costs of liquefied hydrogen.

It is perhaps these three areas where ammonia has some interesting potential as a fuel for the future.

How green could ammonia be?

Ammonia could absolutely be zero-carbon. If you can make green hydrogen (either with green electricity or with carbon capture and storage) and green electricity, you can make green ammonia. You just need a lot of green energy. I do think the focus now should be on de-carbonising the world's existing ammonia demand. This would then provide a sound footing for ammonia to potentially be used as a green fuel.

PART 3: USING ENERGY

15. Is the future of travel all-electric?

In 2019 3.4% of car sales in the UK were electric and by September 2020[73] this had grown even further to nearly 7%[74].With further government support and plenty of new models entering the market (175 by the end of 2020) the growth seems set to continue. The UK government has recently moved its date for ending petrol and diesel car sales from 2040 forward to 2030, although they will allow some hybrids to 2035.[75]

The more ambitious forecasts suggest that by 2030 there could be 130 million electric vehicles on the roads globally.[76] To put that number into context, there are around 1.3 billion cars on the road now[77], so even the most ambitious estimates are still only forecasting 10 percent in the next 10 years.

There is a risk that electric vehicles are in themselves seen as green. It is important to recognise that electric vehicles are only as green as the electricity they use to charge.

Can all vehicles go electric?

Electric vehicles make sense for certain applications. Current battery technology means that most electric vehicles remain suited to shorter journeys. However, high-end vehicles like the Tesla Model S with ranges in excess of 300 miles are pushing boundaries.

Charging-infrastructure is an important barrier in take-up. The electric vehicle experience is inherently less flexible than petrol, and without substantial network investment, use will remain limited to second cars or those using vehicles for reliable journeys.

Some of the larger 'fleets' are looking to electric vehicles, like Amazon, which has ordered 100,000 vehicles[78]. It will be interesting to see how they cope with reduced range on traditional vehicles and changes to ways of working.

What about freight?

Freight is unlikely to go fully electric, as current battery technology is not capable of moving large weights over long distances. Even with significant improvements in technology, freight is unlikely to go electric for a long time.

There are some interesting trials of overhead-line technology (like trains)[79] where, instead of having a battery, a 'pantograph' is raised above the lorry to electric wires overhead. These provide an interesting if currently largely unproven route to the decarbonisation of freight.

As with many energy applications the answer will likely lie in hybrid options. Heavy goods vehicles may, for example, use both batteries and hydrogen fuel cells.

What about other forms of transport?

Rail has a clear pathway to decarbonisation through electrification. Existing electrification projects have, however, been dogged by overspends, with scope reduced.[80]

Could electric vehicles become obsolete?

There is the potential for electric vehicle infrastructure and electric vehicles to become obsolete. In the event that an alternative fuel could be found, such as hydrogen, then the benefits of a new technology could effectively leapfrog electrification. In this case it is possible that electric assets could become 'stranded' as users move to more convenient options.

How are we going to charge all those electric cars?

The challenge we are yet to really experience is that of charging millions of electric vehicles. Electric vehicle chargers use a significant amount of electricity, and electric vehicles are only green if the electricity they are using is green itself.

In order for electric vehicle growth to continue, the ability to charge vehicles in the right places is the bigger challenge. The automotive industry has already provided the vehicles – it is now for the infrastructure to follow.

16. The three futures of flight

The future of air travel just does not seem clear – right now there is not a readily available and technically proven 'green' alternative for air travel.

Prior to Covid-19, every minute 84 flights took off somewhere in the world with more than four billion journeys being made by plane a year[81]. This frequent travel burned nearly three hundred million tonnes of jet fuel annually, making up around two percent of global CO_2 emissions[82]. Before Covid-19, global air travel was expected to double over the next 2 decades[83], a projection that although dinted in the short term remains likely, with increasing numbers of individuals able and willing to pay for the luxury.

Recently I've been excited to see small electric planes taking off, such as the Eviation Alice electric aircraft currently under development by Eviation Aircraft of Israel[84]. Using battery technology taken from the automotive sector, these planes have been able to travel reasonable distances on one charge – and carrying a couple of a passengers. There have also been some bold statements about having electrically powered large aircraft by the end of the decade. These innovation steps, however, are nowhere near to decarbonising a long- (or even a short-) haul flight.

The fundamental challenge of decarbonising flight is the energy density of the storage. The physics of batteries seem to work for smaller applications but not necessarily for bigger aircraft. Currently, lithium-ion batteries can store around 250Wh per Kg, which is 30 times less dense than jet fuel[85]. So, the weight of the batteries ends up limiting the ability of larger planes to even get themselves off the ground – never mind carry a payload. Research has suggested that for battery-powered aircraft to work, the energy density would need to be nearer to 800Wh/Kg - nearly triple that of the best available technology today. At the current rate of technology improvement, this kind of energy density is not going to be available until well after 2050.

Even if the improved energy density could be achieved, a huge challenge remains – that of charging large airliners on the ground in a time frame acceptable to carriers and airports alike. To put it into context, a Boeing 747 needs around 60 megawatts of power for cruising[86]. A Boeing 747 currently has a turnaround time of 150 minutes, and a 737 for budget airlines like Ryanair 25 minutes.[87] The infrastructure required to be able to charge such a large battery so quickly would be quite incredible (never mind charging 1000 flights a day at Heathrow.[88]). Even if the energy density challenge can be resolved, there is still a long way to go to be able to charge the aircraft once it is on the ground.

Finally, biofuels have some potential; however, they are unlikely to be game changing. Currently, biofuels supply about 0.1% of global aviation fuel, and despite some early enthusiasm they have not seen high levels to growth.[89]

Looking to the next decade: what does the above mean for air travel?

The development of smaller emissions electric aeroplanes may mean we see an explosion of smaller regional airports with pilotless air taxis. The shorter available distances will mean that short-hop aviation may become a real thing. Also, smaller planes will be able to fly at lower altitudes avoiding congestion. And with weight becoming critical in range, removing the pilot starts to make sense – accelerating autonomous flying to enhance the economic case.

Electricity demand at transport hubs will become an increasing challenge. With an increase in electric vehicles, the demand on local electricity infrastructure will increase. Heathrow, for example has 51,000 car parking spaces.[90] Just charging that many vehicles will take a huge reinforcement of electricity provision. Assuming slow chargers (three kilowatts) and about one-tenth of vehicles on charge at any one time, you would need a 15-megawatt connection just for the vehicles – never mind the additional load if you wanted to start charging giant airliners (in quick turnaround times) as well. With smaller electric planes there will be no need to use giant hub infrastructure like now – operators will be able to move to cheaper smaller airports where charging provision is more easily provided (from local solar for example).

For larger aircraft, alternative fuels such as hydrogen with higher energy densities may make more sense than electrification, as the energy density of these fuels will make such a switchover more logical. The development of regional hydrogen centres presents opportunities globally, however. As of right now no one is commercially flying on hydrogen. There is still a long way to go.

In conclusion, current available technology means the only way to reduce carbon emissions is to fly less. There just is not a clear enough pathway to lower or zero-carbon flight. Smaller electric planes may supplement existing routes currently served by smaller aircraft such as the Skybus to the Isles of Scilly in the United Kingdom. Without significant regulatory intervention and some big technological leaps, they will not replace our existing hydrocarbon hungry fleet any time soon.

17. Aviation – smaller and spread out?

Prior to Covid-19, the aviation sector was pegged for year-on-year growth. Historically, it has been a sector driven by economies of scale with capacity being concentrated on huge hubs with larger and larger aircraft. Recently, fuel costs have driven airlines away from epically sized jets to more fuel efficient and smaller models; however, airport hubs just seem to grow and grow.

Whenever I look to travel, I am always amazed by the amount of time wasted in larger transport hubs. The length of a short flight can typically be expanded by double or even triple, as time is taken up at either end in traffic, parking, getting through security and waiting for baggage!

Does the energy transition and digitalisation change the way we need to think about aviation?

Future fuels push us to smaller, lighter aircraft.

There are several alternative fuel pathways for flight which are currently getting most attention, two of which, hydrogen and batteries, come with energy density challenges. Batteries still weigh 30 times more than aviation fuel for a comparative amount of energy stored, and hydrogen needs to be chilled and liquified to provide anywhere near enough energy in a reasonable volume to fly an aircraft.

Currently, the largest electric plane called the E-caravan weighs in at about 4 tonnes and can carry about nine passengers for about 100 miles[91]. The challenge of energy density in the batteries remains huge for electric aircraft – not least because current regulations require planes to hold a substantial amount of energy in reserve in case of issues.

The largest hydrogen powered plane – a six-seater aircraft - recently took off in the UK, built by ZeroAvia using hydrogen fuel cells[92]. The company plans to do flights of 250 to 300 miles, and even more recently Airbus has revealed a 'concept' aircraft, which looks similar to what we fly in now – but something like that taking off feels a long way away[93].

Both technologies feel like something that is going to work well for smaller aircraft – and not super jumbos.

Preparing airports for hydrogen or batteries

Much of the focus is around getting aircraft in the air... and not how we would manage the infrastructure to scale thousands of aircraft regularly taking off and landing. Scaling matters when it comes to air travel; pre-Covid-19, 40 million flights took off a year![94] To achieve that epic scale takes some incredible complex and integrated infrastructure on the ground.

Thinking about our two competing technologies, what does that mean?

Electrification: Efficiency in aviation is all about turnaround time. Aircraft sitting on the ground are not earning money. With battery-powered aircraft, airports will need staggering amounts of electricity import capacity. To back this up the electricity system would need to be capable of supplying this power when it was needed – and in a green way!

Hydrogen: Liquid hydrogen in large volumes is quite different from aviation fuel. For a start it needs to be cooled to -259 degrees Celsius! That makes shipping, storing it and loading it on to aircraft quite challenging.

Do not ignore digital disruption

Secondly, digitalisation starts to make things interesting. Pilotless aircraft are certainly technically possible whether that be remote-controlled or even potentially fully automated. Swarms of aircraft become possible as aircraft communicate directly with each other to find the optimum and safest route.

The case for decentralised aviation

In the UK there are literally hundreds of small airfields. In fact, there are so many it is difficult to quantify them. Additionally, the number of potential 'airports' explodes if you start considering the potential for aircraft like the Osprey, which can take-off vertically[95].

These present a tantalising prospect for flight. What if by spreading out the problem we have an opportunity to completely turn on its head the existing flying experience?

Energy Provision: Particularly electrification presents huge challenges if demand is focused on a single point. Spreading the demand around reduces the size of infrastructure required.

Doorstep to Doorstep: Airports are frustrating places, causing a huge concentration of activity in one place. Even the slickest in the world still add a huge amount of time onto any journey.

Convenience: To enable economies of scale we travel to an airport (sometimes hours of driving). Local airfield s could open a new level of convenience.

Experience: Smaller aircraft and airfields open the potential for bespoke experiences. Whereas now we are pushed into vanilla offerings, smaller aircraft completely open opportunities for personalisation.

So is decentralised aviation the future?

I think realistically for long haul you will still need the big hubs (and long runways). But for shorter, greener trips there is real potential for aviation to become more like Uber than it is now. If something is greener, cheaper and convenient, it is going to be difficult to compete with!

18. The carbon footprint of space travel

Who could not fail to have been wowed by the recent, incredible, crewed Space X launch on the 30th of May? Two astronauts successfully left the Earth's atmosphere to dock with the international space station on a semi-reusable rocket. This was clearly an incredible achievement and returned the USA to the forefront of space travel. In addition, this significant step moves us closer to manned flights to Mars, which I believe will highly likely happen within the next decade. I love the idea of NASA outsourcing the 'easy stuff' so that they can focus on the bigger prize of Mars!

Watching the launch got me thinking about the carbon footprint and environmental impact of shifting the 12-tonne Falcon 9 and Dragon capsule 400 kilometres into space to meet the International Space Station (ISS). I was surprised how difficult it is to answer the carbon footprint question – and more worryingly how dubious the maths was where people have had a go[96]. I have pulled together various numbers from across the internet to try and get a feel for the number - my calculations are below if you want to see my underlying logic.

Calculating the carbon footprint

The Falcon 9 rocket is powered by nine Merlin engines. The Merlin engines generate about 1.7 million pounds of thrust at full power, consuming a mix of super-chilled kerosene and cryogenic liquid-oxygen propellants. Around 155 tonnes of the cooled liquid kerosene are consumed during a launch, along with 362 tonnes of liquid oxygen. That is a lot of fuel sitting right underneath our two astronauts. Not only is high-grade aviation fuel being burned, but also a lot of oxygen is being used up in the combustion process. So, what is the carbon footprint of the launch?

- Kerosene has a carbon intensity of 3Kg of carbon per Kg of Kerosene[97]. So the carbon generated from the kerosene used in the launch is 465 tonnes.
- The oxygen used is produced from a cryogenic process which uses electricity to chill air to release the oxygen. Assuming the storage and transportation is relatively efficient and grid electricity is used to produce the oxygen, then the carbon emitted in producing the oxygen is a further 650 tonnes.

Therefore, the total carbon footprint from the kerosene and oxygen is around 1115 tonnes, the annual carbon footprint of 278 average world citizens. In all honesty, I would have expected it to be far greater.

There is an opportunity for the oxygen to be made using zero-carbon electricity - but given no one is shouting about it, I doubt this is happening (I would happily be corrected!).

Comparing that to conventional flight, a Boeing 747 burns about 4 litres of fuel a second; flying from London to New York in total it uses around 70 tonnes with a carbon footprint of around 210 Tonnes of carbon each way. Comparing that to our launch, then, we are using around the equivalent of five return transatlantic flights.

Another measure is emissions per passenger/per km travelled – which for the recent trip to the ISS of just two astronauts is about 700kg/km (I have assumed 400km each way with no fuel burned on the return). That compares to 0.133kg/km for domestic flight or 177kg/km for car travel[98]. This improves significantly once the Dragon capsule has a full complement of seven astronauts.

It is silly to compare space travel with rail and air; however, it does show just how much comparative energy is being used and consequently how much carbon emitted.

Other considerations

There are some other interesting impacts of space travel regarding where the emissions happen - for example, soot in the upper atmosphere and depletion of the ozone layer. I have not gone into these here as they are super-complex and there does not seem to be too much clarity in the science on the impact. They are, however, becoming increasingly important as the number of rocket launches increases.

Conclusion

Annually, on a global basis, there are around 100 space launches a year – however, with space tourism and increasing numbers of satellite launches, this is pegged to rise to well over 1000[99]. If we take the carbon number for our launch, then we get to a carbon footprint for space travel of around 3.1 million tonnes in only a few years' time - along with damage to the ozone layer in addition to soot in the upper atmosphere. As space travel expands and Mars looks increasingly possible, it will become more important to manage the footprint here on earth of our aspirations to explore our solar system.

Some Maths!

There is not a lot of data freely available on the carbon intensity of liquid oxygen. I have taken an example of a 300kW oxygen plant using 300kw of electricity to produce 2 tonnes in a day[100]. So to produce 1 tonne of liquid oxygen you need about 3.6MWHr of electricity! To produce the 362 Tonnes of liquid oxygen needed for the launch, therefore, you need 1300MWHr or electricity. The average grid carbon intensity in the US is 0.5 tonnes per MWHr. Therefore, the carbon generated in producing the oxygen for launch is about 650 tonnes.

19. The four Cs of the energy transition

When looking at the energy transition I believe there are four lenses through which to make energy decisions. We are moving from linear decision-making in an age of energy abundance into a multi-angled challenge which constantly evolves over time.

Carbon intensity - what is our carbon footprint?

Carbon intensity is a complex challenge which is not only hard to nail down now but is even more difficult when you project into the future.

Energy problems need input and outputs. In the case of electricity, the carbon intensity of those inputs can change each half-hour, never mind five years into the future. Since 1990 the carbon intensity of the electricity grid has fallen by 38%[101] and continues to fall in all future energy scenarios I have seen. I do think a lot of these scenarios tend to understate the impact of a huge shift to electrification and the capability of renewable sources to respond to rapidly escalating demand.

That said, gas networks have their own plan for reducing the carbon intensity of the gas in the pipes through bio-methane and hydrogen mixing so the future may not be all-electric.

Capacity – can we get to the energy we need?

Even if we can get an abundance of low carbon energy through hydrogen or renewable electricity, the next challenge comes in getting it to the point of use.

There are not insubstantial challenges with local and national grid capacity in several energy scenarios. These directly impact both individuals and companies seeking to make energy decisions.

That means at a residential level not everyone on a street can have a 20-kilowatt fast electric car charger. If delivery companies quickly electrify their fleets, the grid connections needed to provide even overnight charging will be phenomenal.

Heating UK homes at peak requires about 360GW[102] of instantaneous energy (for comparison current 'peak' electricity demand is about 50GW). Any shift to electrification of heating will massively increase demand on both national and local grids. Even hydrogen may not present the perfect solution – 'green' hydrogen uses electricity to generate it, and huge volumes of storage would be required to enable enough to be there on super-cold days.

Cost – how much will it cost us?

Carbon and capacity both come at a cost. Resilient, zero-carbon options do not always come cheaply. With so many alternative options, delivering the lowest instantaneous and lifetime cost presents a complex and diverse challenge.

Geography is also likely to become far more important. With hydrogen deploying locally and grid re-enforcement relying on substantial infrastructure investment, it may simply be that the costs of energy are influenced equally by where you live or where a company is situated - even down to a parish level in the UK. A little like the two million off-gas grid properties in the UK, who currently have no choice but to opt for more costly liquid petroleum gas, we could see differences depending on hydrogen availability or even local electricity grid capacity.

Cooperation - how can we make this happen?

Finally, and most importantly, the answer to the above three challenges is in cooperation. We simply must look to our neighbours to make better use of what we do have. There are plenty of fantastic examples out there; we are just going to have to work harder and think differently to get there.

- Whole-system planning and local energy systems across zones – enabling capacity to be managed at a more local level.
- Using neighbours' waste heat (data centres, sewage, air conditioning) to optimise performance of heat pumps.
- Understanding local demands on networks - such as electric vehicle charging prior to requesting giant grid connections.
- Shared storage through heat networks enabling reduced demands on the system and energy sharing.

Cooperation increases the number of no-regrets decisions. Energy saving and energy sharing typically always makes sense - and finding ways to limit demand and peak capacity requirements reduces cost for everyone.

This type of thinking takes an unprecedented level of coordination at the local, regional and national levels. Success comes in a constant open dialogue about future and current needs – and maybe giving up a little control for the greater good. The ultimate technology split is still a long way from being decided, but what is clear is that **carbon, capacity, cost and co-operation** are the cornerstones of any decision-making around energy systems.

20. Fifth-generation heat

Heat networks are an interesting proposition for decarbonisation. Government forecasts suggest around 18% of heat demand could be provided through some form of heat network by 2050[103]. Heat-network technology is progressing rapidly with the emergence of a new fifth-generation approach, which certainly requires consideration.

What do we really mean by fifth generation?

The fundamental premise of fifth-generation heat networks is that distribution temperatures drop substantially to around 25 degrees (typically fourth-generation systems run at around 50 to 60 degrees).

At this temperature the system integrates with heat pumps enabling energy sharing – for example, using the 'low grade' heat energy emitted from cooling systems, which can then be redistributed round the network.

To achieve high efficiencies, data is used to constantly optimise the total system – modifying the network temperatures and flows to provide the optimised position for all users. Whereas existing networks may have summer and winter temperature settings, the fifth generation constantly adapts to deliver the optimum temperature, driving the coefficient of performance (how efficiently heat pumps convert electricity into cooling or heating) across the system.

Fifth-generation heat networks have some unexpected and positive consequences over other heat de-carbonisation options. These matter when it comes to building future smart cities, most notably in freeing up roof space to live in (no more big cooling rejection units from air source heat pumps) and ensuring that local cooling effects do not make our city centres more than a little chilly. They also do not have the local emissions issues which burning hydrogen in domestic gas boilers will have.

I did not even know we had been through third and fourth generation. What did I miss?

The trend in heat networks has been to reduce the temperature of the water in the system. Older networks use steam to transmit heat, with more recent systems using water at around 80 degrees. The thinking of academia and industry has recently shifted towards fully integrated heat systems with temperatures as low as 40 degrees in networks.

In practice, fourth and fifth generation networks are likely to merge. Unlike your phone, you do not get a little logo in the top left corner of your screen saying you are using fifth- generation heat. Fifth generation is something that can happen quietly in the background with the end-user not really knowing.

What is happening to make fifth-generation heating technology mainstream?

There remain some significant challenges with fifth-generation technology. How you deliver a system for the 'greater good' and share the carbon and cost benefits will present some interesting challenges.

There are several case studies in operation now which are showing some interesting performance. One is the research being led by South Bank University (amongst others) around the Bunhill[104] cluster in London, which features some great examples of using 'low grade' heat – the London underground, sewage and cable tunnels are already being looked into and the potential is out there.

There is also some interesting research from the likes of Lot-Net, where universities are working together to deliver some fascinating insight into how fifth-generation heat networks can and will work.

What will need to happen to make it work?

Collaboration is key to fifth-generation networks. Parties need share the carbon savings and produce a system where everyone benefits. Opportunities for energy- sharing exist in many places, and it is only through open dialogue that these can be turned into reality.

Also, systems need to be designed 'future proof'. Lower temperature networks and heat pumps need systems that rely on much lower temperatures. A good example is using underfloor heating instead of radiators (or higher-capacity radiators). Some simple choices like taps and shower heads with a low temperature differential can make a huge difference down the line.

The future of energy

PART 4: THE FUTURE

PLEASE

STAY SAFE
1.5M APART

THANK YOU

21. Does Covid-19 change everything?

Two of my favourite books are *The Power of Habit* and *The Chimp Paradox*. In both these books the authors take you through the power of habits on changing mindset, performance and health, and how you can consciously manage your actions and feelings through building up good habits.

Habits become super-interesting when you think about them in the wider context of Covid-19. Since March 2020 we have all to some degree had our worlds turned on their heads. Habits we have carefully maintained (both good and bad) were deconstructed overnight - our morning Starbucks, monthly visits to parents, or quick weekends in Paris all disappeared. As we moved into lockdown, we formed new habits, some good, some bad. The news has been filled with stories of increased demand for sewing machines and exercise bikes, and on the flip side news of increased drinking and screen time, amongst many others.

At the macro level, changes in our activities are clearly seen in energy consumption. Globally oil and electricity demand have dropped during the Covid-19 crisis. The IEA is seeing a reduction in global electricity demand of 2.5% in Q1 of 2020[105] and a staggering 57% reduction in global oil demand – predominantly driven by a huge reduction in transport. 65% of the global aviation fleet sat idle in April 2020![106] Rail travel was particularly hard-hit as workers abandoned city centres en masse, leaving only 17% of workers holding the fort at their desks in August[107]. Microsoft teams, rarely used before lockdown by most, became a staple part of office life – leaping by 70% in just the first month of lockdown to 75m users globally[108].

The habit of commuting

> *'For 20 years I have got up at 6 am to sit in traffic to get to work just to sit at my desk and check email – I will never do that again.'*
> Posted on LinkedIn August 2020

A big impact in the UK of the lockdown has been the near desertion of urban centres. With some major employers saying they have no intention of returning their staff to city centres,[109] and with employees still staying away –only around 17% of staff were back at their desks in August - ,[110] the slow demise of retail has been a known trend; however, city centres had been able to replace retail units with offices and hotels investing in urban areas like never before.

Once people no longer need to be together to be productive, there is a huge paradigm shift. Energy-hungry cities are replaced by more spread-out demand as people change the way they work and travel. For a long time, a top-down energy industry has focused on feeding power-hungry urban centres for five days a week. If we disperse more then decentralised energy becomes super interesting. Local production, local storage and local distribution become key.

The habit of global travel

Another stark image of the impact of Covid-19 has been airplanes lined up at storage sites[111]. 65% of the global fleet was in storage by mid-April 2020, and major global airports have sat empty.

The ease and relative low cost of air travel have made us build up travelling habits. For decades we have learned to enjoy the buzz of planning a weekend escape. We have learned that a successful meeting needs to be face-to-face. In fact, the travel industry feeds the habit with reward programs and status — frequent- flyer programs aren't there because airlines are nice; they are there because they tap into some of our most basic needs and desires (belonging, status, gathering). I will guarantee that once Covid-19 restrictions lift, airlines will be pushing their rewards program hard to energise their most regular customers' old habits! Especially as prior to lockdown, 15% of flyers took 70% of the flights![112]

I believe the impact on air travel will be there, but it will not be as stark as some forecasters are predicting. That is why we need to look for high-density fuels— like synthetic aviation fuel - to get to decarbonised air travel.

What next for energy post Covid-19?

As time progresses, habits become harder to shift – working from home becomes the norm as does online shopping or a trip to Cornwall instead of the Canaries! Energy as a sector is already going through an epic transition and Covid-19 has both accelerated this and delayed it. What we do not yet know is whether there is a significant adjustment in human behaviour to take account of or just a short-term blip in a longer inevitable trend.

There are many sectors impacted by changing habits, and each of these in large or small ways will impact energy – more online shopping means more vans on the road, more time at home means increased domestic heating demand, amongst many other examples.

What matters now is how these habits stick. Which ones will spring back instantly to old ways and which will

stay with us forever? The next six months will be critical in defining the future and how these already slowly embedding routines remain with us into the next decade.

22. How can cities take control of the energy transition?

The energy transition presents cities with unprecedented challenges in planning for the future. Electric vehicles will place an exponential demand on electricity infrastructure, gas is being phased out, and we need increasingly stringent controls on local emissions, particularly in urban centres. Simultaneously, governments are introducing an array of regulations and reporting requirements along with an equivalent and similarly confusing selection of subsidies.

Where are we today?

Cities of all sizes need to draw a line in the sand and assess where they are today: gauge their current consumption, which infrastructure assets already exist, how energy is already used.

Taking this to the next level, it gets interesting -- what is the capacity of the existing systems, how many parking spaces are there that are sustainable for future electric vehicles, how do traffic flows look now, how long are people spending in the central business district? Importantly, capacity maps for all the main utilities are essential – identifying bottlenecks now can reap huge benefits later.

This is where data comes in. Amazing datasets exist of city energy consumption, people movements, traffic flows, satellite heat maps, underground infrastructure. Bringing these together can be bewildering but will help construct a "digital twin" over time – enabling exponentially better planning.

Consider the city's direction of travel

Cities around the world are declaring climate emergencies as well as seeking to address local pollution. This is a great time to take stock and establish a clear direction of travel.

Firstly, electric vehicles need to form part of any plan – soon a robust charge-point network will be essential in all urban centres. A fast charge point uses 22 kilowatts, so a handful of charge points in one location can quickly push local energy infrastructure to its limits. There is also a temptation to ignore residents without dedicated parking spaces. Currently, a lack of available infrastructure for those without their own drive or dedicated space effectively locks many citizens out of joining the electric vehicle movement.

Cities typically have huge asset bases themselves, with hundreds of energy-hungry buildings. A clear leadership strategy is needed for building fabric, energy provision and consumption. Realistically, natural gas is going to be around for a while; however, the city may want to decarbonise quicker, it may have ageing assets that are due for a replacement, or it may need a little more resilience – in which case some decisions may need to be made sooner rather than later.

Now is the time to be thinking about upping electricity-distribution capacity – ready for rapid expansion both for electric vehicle charging and heat pumps across urban centres. This change is happening as we speak, and demand will likely rapidly outstrip supply.

Look to your neighbours

The energy transition is all about collaboration. Using the available data sets mentioned above, opportunities to share energy can easily be identified. Take a look around you for collaboration opportunities – do neighbours have excess photovoltaic panels or spare heat (such as data centres which eject huge amounts of heat in the process of cooling computer servers)? Alternatively, do neighbouring businesses and government buildings need to charge electrical vehicles overnight (for example, delivery vans) whereas commuters need to charge them during the day?

The future energy world is all about collaboration – network pinch points, particularly electrical ones, are going to drive us together into local problem solving. City representatives can play a huge part in facilitating and even leading that collaboration.

Use energy scenarios

Consider the cities strategy; for example, what are your plans for the next 10, 25, 50 years? Even 10 years feels an awfully long way away, so scenarios can help. Cities are in a unique position, so they can play the long game. Unlike businesses with shareholders to satisfy and a need to chase return on investment, cities seek different objectives.

Energy trends around the electrification of heat, electric vehicles and digitalisation are clear – there is a lot of talk about hydrogen but realistically this is beyond the planning horizon for most of us. That said, for cities, now may be the time to make systems at least 'hydrogen ready' in some way.

Take small steps now in the direction you want to go

The thousand-mile journey really does start with a single step. There are some no-regrets decisions. Solar photovoltaics will generate carbon-free electricity for 25 years; energy efficiency always makes long-term sense and should always be the 'first fuel' in this conversation. Electric vehicles are only going to grow in popularity and energy demand. Local interventions to lower nitrogen oxide and particulate emissions are simply the right thing to do from an environmental, health and safety, even from a child-protection perspective.

For complex, interdependent infrastructure, the array of challenges can seem bewildering. Some cities have made clear steps – enforcing ultra-low emissions zones, supporting heat-network infrastructure, replacing diesel busses with electric alternatives. I have been particularly impressed with some of the bold moves made to address infrastructure challenges in cities like London, Copenhagen and Bath.

Finally, cities need to look for support and need not wait for a clear plan. As I said at the beginning of this chapter, there are a sometimes-bewildering array of subsidies out there. Globally, cities are doing some incredible things; the C40 cities are a fantastic example.

23. Energy efficiency – the 2-billion-tonne challenge

We simply cannot wait for 'silver bullet' technologies to take away our problems when we have the technology now to do something about it. By building an efficient, low temperature, digitally linked system, significant decarbonisation of heat is achievable.

To accelerate the decarbonisation of heat we must:

- Ensure new buildings are highly efficient and future-ready, and accelerate efforts to improve existing housing stock.
- Focus on digitalising the energy system to make it truly 'smart'.
- Grow no-regrets options such as efficiency improvements, heat networks and waste-heat recovery.

So where are we now?

About half of energy consumed is used for heating, and half of that is used for heating buildings and hot water, contributing 22% of UK greenhouse emissions (100 million tonnes of CO_2 per year).[113]

There are about 30 million homes in the UK of which 83% rely on natural gas for heating[114]. It is interesting to note that in 1970 it was only 43%. Gas heating has grown in a world of abundance and low costs. Leaving us now with a legacy of poorly insulated homes contributing over 5 tonnes of carbon per home per year to emissions - where the technology exists to easily halve these emissions today (so over 2 billion tonnes of CO_2 will be unnecessarily emitted between now and 2050 if we carry on doing what we are doing now). And that is just using the domestic figures, never mind the commercial ones.

To date, significant steps have been made in reducing the carbon intensity of power; in fact, three-quarters of the reductions since 2012 have come from power[115]. The UK system lasted a week without generating any electricity from coal recently demonstrates just how far we have come.

These improvements mask little improvement in heat and transport. With electrification of transport progressing (although too slowly), heat remains the final frontier of the epic energy transition which we are currently experiencing.

We now need to urgently address the challenge of decarbonising heat, which will require a transformational change to the energy system. Through investments in energy efficiency, digitalisation and networks, we can now make a change like that achieved in the electricity generation industry in the preceding decade. Through making strong choices now, we can (and must) achieve a step-change in decarbonisation like that achieved in power.

Energy efficiency

So where do we start? Energy efficiency matters. Before we discuss how to generate heat in different ways, it is essential to look at how we can use less and how we can future-proof systems to be able to work well with future technologies.

House efficiency is measured using an alphabetical rating from A to G. A typical 'A'-rated house has energy costs of around 400 pounds a year compared to over 2000 for a 'F'- or 'G'-rated dwelling. Most properties in the UK are rated 'D' with consumption around double of an 'A'-rated building[116]. Some recent analysis by the smart-thermostat manufacturer 'Tado' showed UK houses performing on average the worst in Western Europe with an average drop of 2.5°C at a zero ambient temperature over 5 hours – compared to around 1 to 1.5°C in other countries.[117]

By 2050 95% of homes will need to be 'A'-rated – that means improving the energy efficiency of 15.5m homes between now and then. That is an average of half a million a year.

The real key to this is that heating properties inefficiently typically uses higher temperatures. The recent study by the consultant Ramboll on converting the energy system of the town of Cowdenbeath to different systems highlighted that investing in energy efficiency and running at lower temperatures had benefits in all scenarios[118].

New funding mechanisms are required now to incentivise rapid investment in energy efficiency.

New-build houses

New-build has an important role to play. Each year a further one 143 thousand new houses and flats are built (which could easily rise to 300 thousand)[119]. New buildings should be built with the highest levels of efficiency. Banning domestic gas boilers from 2025 is a strong step in the right direction and will encourage investment in alternatives – with further investment bringing down the costs of technologies such as heat pumps, similar to what happened to offshore wind, where through industry and government-backing, costs were substantially reduced as the industry grew.

Establishing high efficiency, low temperature systems in these properties will pay dividends later.

Low-carbon heat networks

Significant savings in energy consumption can be achieved where heating and cooling are situated in proximity. These seek to create a 'breathing' network where heat is used again and again as it passes through the city environment. Academic research such as Lot-Net is making excellent progress in the area of understanding how these 'breathing' networks can be established.

One cornerstone for these networks to operate well is the integration with the end-consumers, where good building insulation and low temperature systems smooth peak demands and lower overall temperatures. These enable systems to tap into 'waste' heat such as sewage, rivers and tunnels using heat pumps with extremely high 'co-efficient of performance' values. A high co-efficient of performance means lots of heat is produced for extraordinarily little electrical input.

Heat networks also provide a significant storage opportunity. In times of high demand, networks can store significant volumes of energy as well as potential, using 'phase change' storage. This can make a huge inroad into the challenge of smoothing the peak energy demand.

What matters is that networks become smart – providing grid balancing through utilising their immense storage potential.

Heat pumps & hydrogen

Heat pumps have the potential to provide a replacement technology to existing gas boilers. They can play an important role in transitioning to a low carbon system. Hybrid heat pumps, which use gas to meet the occasional 'peak' heat demand periods, offer a potential no-regrets option. These are a good option for retrofit. It is, however, worth highlighting that installations need to have two key features. Firstly, the heating system needs to be improved to operate at lower temperatures, and secondly, the system needs to be 'smart' to limit peak demand.

Evolution of the new build-market is extremely relevant to lowering costs and preparing for mass rollout of heat pumps into the retrofit market.

The government's future-homes standard will play a key role in supporting the deployment of heat pumps at scale in new-build, providing confidence to the supply chain to invest and innovate to meet this demand.

Hydrogen remains a high-cost option, requiring either large volumes of extremely low-carbon electricity or carbon capture and storage. Of the options that can be deployed today at scale, it will not be viable for some years to come. Should we wait for hydrogen at a time when action is required today? In 10 years' time when the technology might be ready, we will have already unnecessarily emitted well over half a billion tonnes of carbon from our housing stock.

What will the system look like to support the transition?

Heating is distinctly seasonal, with peak demands being 5 times higher than peak electricity demand. Heating needs around three hundred gigawatts of energy to be delivered at peak. For reference, the electricity grid can deliver around 60 gigawatts at peak. Whether you use heat pumps or hydrogen, this is a challenge to overcome. Technologies such as batteries, storage and smoother consumption can make a huge difference.

Clearly the electrification of heat will increase demand significantly over time – but a significant amount can be done to meet this expectation. Through building an efficient, low temperature, digitally linked system, electrification of heat is achievable.

A call to arms

I believe that the challenge of the heat transition can be achieved if we:

- Invest in future-ready, highly efficient buildings which operate with lower demands and temperatures whilst accelerating retrofitting the existing housing stock.
- Support heat networks and other low-temperature systems - those that can utilise sources of waste heat.
- Invest in making the system 'smart' through smarter grids, smart meters and systems to

facilitate time-of-day pricing, load shifting and storage.

To do this we need to act now to incentivise rapid growth in energy efficiency, digitalisation and low-carbon heat sources. With the right environment we can absolutely achieve rapid decarbonisation of heat over the next decade. replicating that achieved in other sectors. We have the technology now to take big chunks out of the 2 billion-tonne challenge – so why wait? The earlier you start the less you emit. A goal of zero by 2050 is great – but we must not miss the clear fact that savings now matter. We can do so much more by acting right away.

24. Will energy still come in vanilla?

Gas and electricity are a pretty 'vanilla' product. Carbon intensity of gas and electricity is typically reported as an annual average, and unless you are a large commercial consumer, kilowatt hours are priced at a flat rate regardless of time of day or even year. Neither of which are reflective of the true nature of the energy flowing through gas pipes and electricity cables, in which the carbon intensities and costs are constantly changing based on millions of complex inputs and outputs to the system.

Carbon intensity (the ice cream)

Looking at carbon intensity of the grid we can see just how much this varies during the day[120]. Even where you are in the country has an impact.

Where a supplier buys its gas also matters. Statoil's Sleipner gas field off the coast of Norway limits CO_2 from production by capturing carbon in the rocks below the sea bed, whereas cooled liquid natural gas (using a lot of energy and shipped from a long way away) will have a much higher carbon intensity. So where the gas is bought from really does matter.

The ability to achieve the best carbon outcome in any given period is impossible to calculate in the moment. There are simply too many variables. However, in the future, technology should help us dynamically choose how green we want to be in any given moment.

Price (the cone)

Having decided on how green I want my carbon, I can then look at price. If I have photovoltaics on site, then this is likely to be my best option. So I need to prioritise when it is sunny. But what if I have a battery and I am expecting prices to go up later in the day? As mentioned above, only large consumers have what is called 'time of day' pricing. With technology there is a view that this will come to all sectors to manage peak demand.

For gas there are peaks in the network where gas storage is used (at a cost to suppliers) to prop up demand (equivalent to about 300 gigawatts of energy at peak). Again, currently there is not a lot of time-of-day pricing for gas; however, it is likely to increase.

So do I want to use my energy at the most cost-effective times, or do I not really mind and just want convenience? Could I cap my costs so at least my system avoids winter pricing extremes?

Customer service (chocolate or strawberry sauce and a flake?)

So how do I want to be served? Do I want to talk to someone directly or am I happy to just be served online talking to a chat bot (is artificial intelligence so good I do not even know the difference?). And then there are the extra bits. What else comes with your energy – such as smart controls, boiler maintenance or other exciting additions?

What does this mean for the future of energy?

In the future, artificial intelligence and smart controls can start to make a lot of the above decisions for us. Just as you can select the type of route you want in a sat-nav, a smart system links together consumption, local generation and supply data to provide the optimum outcome based on consumer wishes.

These kinds of considerations apply to both commercial and domestic environments. Economically, investment in energy performance tends to stack up quicker for commercial users; however, domestic users can be quicker on the uptake for the latest trends as decisions are taken in a hugely different way. It is going to be exciting to see how we stack up our energy in the future.

25. Thinking about resilience

At 4:52 pm on Friday 9th August a lightning strike hit a transmission circuit at 4.5 km from the Wymondley substation. It was a normal day in the UK transmission system, but that single bolt of lightning caused significant disruption. Trains were stopped for hours and the country took a significant time to recover.

> 'Prior to 4:52pm on Friday 09 August Great Britain's electricity system was operating as normal. There was some heavy rain and lightning, it was windy and warm – it was not unusual weather for this time of year. Overall, demand for the day was forecast to be like what was experienced on the previous Friday. Around 30% of the generation was from wind, 30% from gas and 20% from Nuclear and 10% from inter-connectors' Report into to the events on Friday 09th August.[121]

The lightning that hit a powerline at 4:42 pm turned off the power to over one million people. Although power was restored within an hour, the impact went on late into the evening and was particularly felt in the rail system where one fleet of trains could not reset themselves when the power returned. This caused significant disruption (and discomfort) for those stuck on the trains waiting to be rescued.

Was it the fault of wind and solar?

Much has been made in the press of the changing nature of the electricity grid and the impact of more generation embedded in the system. In this case, failures at Hornsea 1 wind farm compounded system issues and contributed to the blackout.

Logically, more decentralisation should build in resilience as large, single points of failure disappear. However, it is clear from the August power cuts that these key single points still exist. In such a complex, interdependent system, a small disruption can still have a much broader impact. Although in this case the lightning strike caused a very short-term impact, it was the lack of resilience of certain critical elements to voltage drops which caused the largest effect. Had the trains been able to be restarted by the drivers, for example, the impact would have been significantly reduced.

In thinking about energy resilience, it is important not to use only past events to plan. In many ways the root cause is of less importance than the impact. A quick read of the list of global power outages shows the broad range of root causes which can impact the system, many of which are beyond our control such as:

- **Weather events** – If you look at large blackouts around the world many are caused by large storms, ice build-up on power lines and lightening.

- **Solar storms** – According to scientists, space weather impacting computer systems and electricity grids is inevitable.[122]

- **Technical failure** – Sometimes equipment does fail like in the London Blackouts in 2003[123].

- **Third party intervention** – A digger hits a cable or there's a cyber-attack.

- **Pandemic** – A pandemic prevents energy workers from attending work to maintain the system.

So what can we do?

All the above events (and many more) can cause electrical systems to shut down. You cannot plan and prepare for every eventuality; however, what you can do is test systems for recovery and assess potential impacts. Following the August power cuts, it is important for organisations (and individuals) to consider:

- What is the impact of even a short interruption of key supplies (power, water, gas etc) to our activities and our customers?
- Is there any way to physically test how we would recover in the event of a loss of critical supply?
- Can we look for ways to build natural resilience into our systems? Renewable technology gives us the chance to build in that resilience - for example, through 'local energy systems'.

As the system evolves and further decentralises, considering our own more local systems becomes increasingly important. A decentralised system can be both more and less resilient in different ways, and users need to consider in such a rapidly changing environment how best to mitigate the impact of system disruptions.

26. How will energy look in 2031?

Predictions are fun, aren't they? Deep down you know even your boldest shot will be wrong - but that does not stop us all from trying. With energy decisions lasting decades, attempting to predict the future is an essential art.

Ten years ago Gareth Malone had the Christmas number one with the Military Wives Choir, and David Cameron was in his second year in office. The UK government predicted in its base case energy scenario that in 2021 49.5 TWh of electricity would be generated from coal[124] and 118 TWh from renewable sources, a significant change from their predictions in 2020[125] when they said it would be 75 TWh from coal and 52 TWh from renewable sources… the reality today is less than one terawatt hour from coal and 127 TWh from renewables. In the space of a year the government got surprisingly accurate with its renewable predictions but could not quite give up on coal yet! This epic under-estimate demonstrates just how wildly wrong we can be over a decade. The pace of the demise of coal in the UK has been massively under-forecast year on year –

highlighted by the record-breaking 67-day coal-free generation streak achieved over the summer of 2020[126] and an impressive 29% reduction in emissions overall in the UK over that same period, predominantly achieved through decarbonising electricity generation[127].

Looking back to our predictions a decade ago is important in forming our views on the next ten years as it shows just how quickly things can change when the mix of regulation, technology and market forces are right.

If anything, I believe this gives fire to making bolder predictions in terms of decarbonisation in the future It does, however, make it a little difficult to pin down a view of the future, as trends cannot be relied on due to the influence of tipping points.

The one constant in any future view is a drive to continued carbon reduction. So here we go. I suspect this may be open to a significant amount of debate and challenge.

The decarbonisation of heat will gain speed with substantial consumer take-up of air- and ground-source heat pumps.

Heat currently accounts for half of energy use and consequently a high share of emissions[128]. Heat pump take-up remains sluggish. However, the recent ban on gas boilers is only the start in several regulatory interventions to drive changes in consumer behaviour around heat. I would predict that by 2030 at least one third of houses in the UK will be heated by heat pumps. This will be driven by regulatory intervention as well as changing consumer perception of methane gas as an environmentally damaging option over electricity.

In 2020 the UK government set out a goal of installing 600 thousand heat pumps a year by 2028[129] - an ambitious target but not quite ambitious enough.

In city centres something different will be needed, as there will not be enough space for ground loops, and air source heat pumps will take up much-needed roof space and cause localised cooling – the impact of which could seriously lower local efficiency.

Enhanced insulation and improved efficiency will gain a sense of urgency and pace.

Currently 14 million properties in the UK have an energy efficiency rating of 'D' or less[130]. That basically means they use twice as much energy as a 'A'-rated property. It is important to note that UK properties are some of the worst performing in Europe.[131]

These properties need better insulation (roof, wall and flow), new windows and lower temperature heating systems. To meet climate targets half a million of these need a deep improvement every year between now and 2050. After a slow start getting our heads around this, I would expect at least five million of the 14 million 'D'-rated properties to have been fixed by 2031.

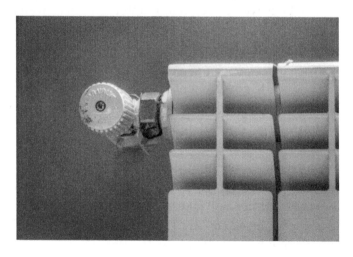

There will be, however, a stubborn core of properties on outdated technology. Even with recent leaps forward I was amazed to see that in the EU there are half a billion radiators without a simple temperature control valve.[132] That's just incredible when you think of the energy saving potential of such a simple change.

The UK government has committed to install 600 thousand heat pumps a year by 2028, which if achieved will make an impressive step towards this target.[133]

Hydrogen will develop in pockets as well as for transport (trains and heavy goods vehicles).

Regional projects like Hy-deploy in the UK will see pockets of intense hydrogen deployment. Whilst in these areas piped hydrogen to properties will be a reality – the slow roll-out will mean that lots of properties electrify with heat pumps before widespread hydrogen becomes a reality. Hydrogen will more likely be used for road haulage and trains, where its high-density energy storage enables it to be the leading low-carbon option. Where and how the hydrogen is made has the potential to become a political issue. With higher electricity costs, using green electricity to make hydrogen will become challenging – pushing producers towards 'blue' or 'brown' hydrogen made from cracking methane with steam reformation (a not so carbon-friendly process unless capture and storage is available).

'Zombie' gas grids increase, pushing costs onto those unable to switch away.

As the electrification (and efficiency improvements) gather pace, the number of people disconnecting from the gas grids will increase. If you take the above numbers for heat pumps and efficiency, then by 2030 there are seven million fewer gas users, and those who remain on the grid are using substantially less. The network costs remain to be paid, pushing up costs on those who cannot disconnect for economic or technical reasons. By 2030 I would expect us to be making a call as to whether to switch off the gas networks in certain high-cost regions.

Tidal barrages – one or two may happen - but the capital costs and local impacts will make further projects difficult.

The idea of generating electricity from capturing the tide remains tempting. These projects have the potential to deliver one-fifth of energy demand[134]. However, the environmental impact of building a tidal lagoon remains significant. With several projects in the pipeline, I believe that one of them must get traction and take off. The lure of zero-carbon energy forever must surely become too great to do nothing.

Batteries – expansion will continue but will be stalled by resource.

Batteries are the ultimate cannibalising technology. The more batteries on the system, the less profit any of them can make. Batteries also have a limiting factor in terms of the minerals used to make them. Already electric car manufacturers are struggling to source enough lithium and cobalt, and these constraints will push up costs.

The current trend is away from decentralised batteries towards larger units at key nodes in the system. I would expect this to continue as local demands increase to support the network.

The system will get smarter (and more local).

Quietly, the system is getting smarter. If you would have said ten years ago, we would be asking 'Alexa' to pop the heating on in the office, we would have said that was something out of Star Trek. But here we are - connected devices are everywhere, and our energy systems are getting quietly smarter. These little bits of inter-connectivity slowly work to improve the system one small step at a time. It may feel slow but as each device interconnects, the ability of the system to make smart choices increases. Consumer demand for smarter systems has the potential to enable a smart energy system as a by-product to other much more fun functionality (such as asking Alexa jokes over breakfast.).

I would expect by 2031 domestic energy to be smart in some form in most properties. Heat pumps will come internet-connected already, and replacement gas boilers will be smart in some form. To meet demand the electricity network will need to get much better at limiting peak demands. Through linked devices much of this is achievable.

That said, resilience remains a concern, with system failures becoming more common due to increasingly poor weather. Local energy systems become more prevalent as a way for communities and organisations to take control of their own energy destiny - moving away from central grids and to locally managed electricity and heat networks, possibly even disconnecting from the centralised systems completely.

Heat networks will become sharing networks, driving collaboration.

Heat networks are predicted to supply 18% of the UK's heat by 2050[135]. My prediction is that they could do more. High-density areas such as city centres have the potential to share energy much more than they do now, taking 'low-grade' heat from sewers, trains, tunnels, data centres and cooling systems to use in heating. Future fifth generation heat networks will use a mixture of technology and lower temperature systems to enable users to feed in and take out energy from the system. Once these systems are established, the benefits to connect will be significant, resulting in a rapid take-up of the technology as networks expand.

By 2030 I believe most city centres will have some form of sharing a 'fifth generation' network or be working towards installing one.

Transport – half of vehicles electric by 2031.

I am going to keep to my convictions and go big here. My prediction is that by around 2030 half of vehicles will be electric, and non-electric cars will be banned from city centres. Recent campaigns by the likes of *The Times* newspaper[136] have highlighted the impact on health from particulates and nitrogen oxides. Recently, the number of cities tightening controls on vehicles has hugely increased, with Bath, Bristol and the City of London taking substantial steps to reduce polluting vehicles in city centres.

I think this will continue and the take-up of electric vehicles will be driven not by concerns over carbon emissions but more by local air quality driving polluting vehicles away from our urban areas.

I also believe that with autonomous vehicles, the rise of the likes of Uber and an increasingly 'rental' based economy, car ownership will drop significantly. Once you can hail a cab reliably for the same cost, why do you need to own your own vehicle (and have the hassle of parking it)? This means that electric vehicles increase in potential as they can take themselves off to charge, with car ownership potentially halving as the need to own your own car disappears.

Air travel will be less common.

Air travel remains a significant challenge. I just do not see a technology yet that can get a lump of an aeroplane into the sky without using a significant volume of hydrocarbons. I can see a world where frequent flying is no longer a badge of honour and instead, frequent flyers are increasingly penalised.

I would expect air travel to have changed significantly by 2031, with greener synthetic fuels being deployed into major transport hubs and a move towards smaller, more efficient aircraft.

Smaller, regional air travel will grow, with electric planes providing regional services.

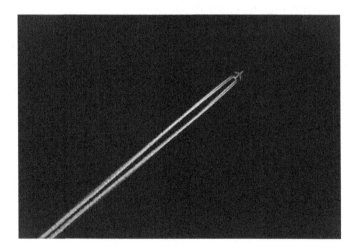

How about a black swan?

Finally, all the above is written from a current view of the world. What is the black-swan event that derails this completely? (Was offshore wind back in 2011 a black swan that was hiding in plain sight?) Nuclear fusion and deep geothermal or exponential development of hydrogen all have the potential to significantly change the playing field. This would, of course, change everything I have said above.

I have recently seen a new term, 'green swan', emerging, referring to foreseeable events caused by climate change!

Time will tell.

Only time will tell if in 2031 the UK will have:

- Five million properties moved from 'D' to 'A' energy performance through deep retrofit.
- Ten million domestic heat pumps installed (ground and air source).
- Two major city centres with a low temperature, fifth generation heat network.
- Hydrogen in the pipes for one or two regions (two million houses).
- Half of vehicles on the roads electric, with a significant reduction in car ownership.
- One or two large tidal lagoons.

- A declining number of gas grids along with an increasing zombie effect hitting costs for the remaining few connected.
- Huge changes in the way we fly.

So what do you think? How far off am I? Have I been bold enough?

I will come back and track these in the 2022 edition.

The future of energy

27. Black (or green) swans

A black-swan event is something that happens which completely changes everything that with hindsight was completely predictable. All our predictions have the potential to be completely derailed by a black-swan event or technology. Technology that has been hiding in plain sight all along. In ten years' time when we look back at where we are today, there will be changes which seem obvious in hindsight, but from where we are sitting today, things are not so clear. Just as offshore wind seemed a technology with potential a decade ago, what will be the offshore wind potential of the decade which begins this year?

Will the dominant technology be an evolution of existing technology?

Battery technology improvements have the potential to be game changing. In 2010 the cost of batteries was around $1000 per kWh[137] dropping to a little over $100 per kWh today. This staggering reduction, if it were to continue, could make battery storage so cheap that electrification of a huge number of applications is possible. At $156 per kWh the technology is not far off-parity with internal-combustion engines. This reduction in storage costs is game-changing and has the potential to push hydrogen technology to the back burner. The ability to store larger amounts of energy makes decentralised solar more viable and pushes us towards 'off grid' properties able to go it alone.

Developments in solar have the potential to be disruptive. Concentrated solar has the potential to provide a route for high temperature processes and enable significant improvement and flexibility in solar generation. The potential to install giant solar farms in sunnier places and transmit electricity long distances also remains a possible extension of already well understood and proven technologies.

Developments in hydrogen are also interesting. Although quite costly now, if hydrogen production followed a similar downward cost trajectory as solar has done, hydrogen could become super-competitive. Emission-free production methods, such as pyrolysis, if scaled up could push hydrogen to the top of the energy leader board.

Is there something completely new?

Nuclear fusion has been a future energy technology for a long time. If scientists are to be believed, nuclear fusion has the potential to provide huge volumes of energy for truly little cost and with little waste (unlike harmful fission by products).

Energy from space is another similarly touted 'infinite' source of energy. The premise is to put solar panels into orbit and then 'beam' down the energy to earth[138]. The limiting factor currently is the cost effectiveness of getting these structures into orbit. There is also, of course, the risk of a James Bond-style laser beam shooting down from space.

The likely 'green' swan

It is likely that the green swan which will define the next decade of energy is already well developed. Just as the rapid expansion in offshore wind seems obvious now in hindsight, will it be heat pumps, hydrogen or even ammonia which define the next ten years? Only time will tell.

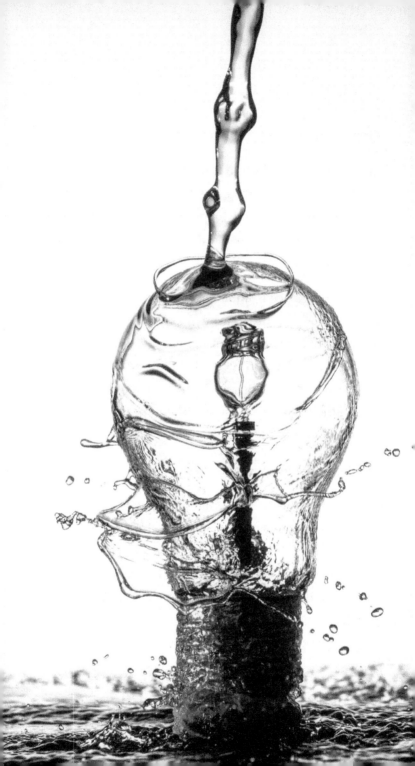

PART 5: CONCLUSION

28. Conclusion

Returning to the *The Future of Energy* in 2021 following one of the most challenging and turbulent years in recent history has been an exciting challenge. The original book was written as a snapshot in time, designed to be kept updated as the topic moved on. No year could better highlight the importance of this philosophy than 2020. Covid-19 has flipped the world on its head, changing the way we work, travel and interact beyond our most extreme expectations.

In energy there is no free lunch. All technologies, be it for generation, distribution or consumption, come with pros and cons, be they cost, carbon or reliability. The temptation to reach for silver-bullet technologies is strong but must be resisted. The media is constantly full of answers which falsely proclaim to be the future. One thing is clear and that is: decision-making must be bold. Technologies have the potential to cannibalise each other – preventing optimum routes from being taken. Failing to recognise the need to phase between alternatives carefully risks leaving the system unstable, costly and unreliable. Different primary energy sources may suit different needs, and it is important that full consideration is given to which technology works best for each application.

Energy is deeply complex and interrelated – it is not just a difficult engineering problem but a multi-dimensional social, political, and economic one. Years of energy abundance have got us to where we are today, with reliable sources of energy and geographically consistent pricing. The future is likely to be much more complex, and as such we must constantly seek to understand the consequences of each new technology not just regarding the cost but also the broader social impact of changes.

I am one of the last to join generation X[139], having been born between the mid-sixties and 1980, and as such have benefited hugely from living in an age of energy abundance and freedom. I have been able to work with different cultures, study abroad and live a life of extremely limited restriction. Along with leaving a world which is safe for my children to live in, I would like them to live rich lives, benefitting from being able to mix with people from other countries, travel easily and safely, and enjoy the freedoms I have enjoyed. I believe that this world is possible, but it is going to take some huge leaps.

Tracking the book and its predictions over 12 months has already demonstrated how quickly change takes place in the field of energy. My still young predictions have already shifted, and what seemed bold a few months ago now does not go far enough. Looking a decade into the future is almost impossible as technologies grow, regulators move to limit carbon and 'green' swans swoop in to disrupt what we thought we knew. However, only through plenty of open dialogue on future pathways will we understand and communicate the impact of the choices that we are making today on both our and our children's lives in the future.

Thank you for taking the time to join me on this journey. I look forward to seeing how energy has developed in the 2022 edition. I would genuinely welcome any comments, both positive and negative, about the book. All feedback is appreciated and contributes towards making the next book better.

PART 6: APPENDICES

Energy units

Energy is typically measured in watts. Unfortunately, one watt is an incredibly small measure and so to make energy numbers useable, engineers use multiples of a thousand watts i.e., Kilo, Mega, Giga and Tera.

To make matters even more confusing we then distinguish between 'instantaneous' energy use and total energy used. Where total is required hours are added to the units for example on energy bills.

Hopefully, the list below helps make this clearer.

Watts (W)

A typical energy saving light bulb uses around 6 watts (6W). Running the light bulb for a day uses 144Wh (which is more commonly described as 0.14 kWh).

Kilowatts (kW): 1,000 Watts (Most commonly used for domestic energy)

A domestic oven uses 3kW. Running the oven for two hours uses 6kWh.

Megawatts (MW): 1,000 Kilowatts (Most commonly used for power generation and distribution).

A gas power station is 450MW. In one hour, the power station generated 450MWh.

Gigawatts (Gw): 1,000 Megawatts (used to describe national demand)

The typical instantaneous demand for electricity in the UK is around 40 gigawatts (GW).

Terawatts (Tw): 1000 Gigawatts (A big number. Used to describe country-level consumption)

Annual UK electricity consumption is 310 Terawatt Hours (TWh)

Measuring greenhouse gases

Carbon intensity (carbon dioxide equivalent)

1 cup of coffee – 21g[140]

Cheeseburger – 4kg[141] (of which 2.6kg is from cattle flatulence and waste!)

1 hour of surgery under anaesthetic - 24Kg[142]

Return flight London to New York – 1.6 Tonnes (8 Tonnes if first class)[143]

Train trip London to Paris via Eurostar – 4.1kg [144]

Average UK citizen – 5.3 Tonnes[145] annually

United Kingdom annual emissions – 435 Million Tonnes[146]

Annual global greenhouse gas emissions: 36 Billion Tonnes[147]

Volumetric measures of carbon dioxide

1.964 kg or CO2 fills a square meter at room temperature.

56kg of CO2 to fill a double decker bus.[148]

Useful resources

The following provide some excellent and regularly updated information:

Bloomberg New Energy Finance: about.bnef.com

Carbon Footprints: www.carbonfootprint.com

Energy Central: www.energycentral.com

Global Carbon Atlas: www.globalcarbonatlas.org

International Energy Agency: www.iea.org

National Grid Future Energy Scenarios:
fes.nationalgrid.com

Shell Energy Scenarios: www.shell.com/energy-and-innovation/the-energy-future/scenarios.html

Sustainable Energy Without the Hot Air:
www.withoutthehotair.co.uk

UK Electricity Grid status:
www.gridwatch.templar.co.uk

World Economic Forum: www.weforum.com

Photo credits

Images are courtesy of the wonderful Unsplash (www.unsplash.com), iStock under licence, or are the author's own.

About the author

John Armstrong is an engineer whose career has spanned the extremes of the energy industry – giving him a front seat on the energy rollercoaster. He began his career constructing oil refineries before moving to work across fossil and renewable electricity generation. More recently John has been leading the growth of decentralised energy and district heating in the UK and now manages energy infrastructure assets.

John lives in Wiltshire near Bath with his wife and two daughters.

References

[1]www.assets.publishing.service.gov.uk/government/uploads/system/uploads/attachment_data/file/61934/national_risk_register.pdf

[2] www.autocar.co.uk/car-news/industry/report-government-ban-new-petrol-and-diesel-car-sales-2030

[3] www.c2es.org/content/international-emissions/

[4] www.globalcarbonatlas.org/en/CO2-emissions
www.statista.com/chart/16292/per-capita-co2-emissions-of-the-largest-economies/

[5]www.carbonbrief.org/analysis-coronavirus-set-to-cause-largest-ever-annual-fall-in-co2-emissions

[6]www.gov.uk/government/news/uk-becomes-first-major-economy-to-pass-net-zero-emissions-law

[7] www.gov.uk/government/news/uk-becomes-first-major-economy-to-pass-net-zero-emissions-law

[8]www.c2es.org/content/international-emissions/

[9] www.mckinsey.com/industries/oil-and-gas/our-insights/global-energy-perspective-2019

[10]gwec.net/global-wind-report-2019/

[11]www.gov.uk/government/collections/energy-trends

[12]www.theguardian.com/business/2020/jun/25/renewable-energy-breaks-uk-record-in-first-quarter-of-2020

[13]www.woodmac.com/press-releases/global-wind-power-capacity-to-grow-by-112-over-next-10-years/

[14]www.theguardian.com/environment/2020/aug/05/china-poised-to-power-huge-growth-in-global-offshore-wind-energy

[15]www.gov.uk/government/news/pm-outlines-his-ten-point-plan-for-a-green-industrial-revolution-for-250000-jobs

[16]www.gridwatch.templar.co.uk

[17]www.carbonbrief.org/analysis-record-low-uk-offshore-wind-

cheaper-than-existing-gas-plants-by-2023

[18]www.newscientist.com/lastword/mg24332461-400-what-is-the-carbon-payback-period-for-a-wind-turbine/

[19]www.research.ed.ac.uk/portal/files/19730353/Executive_Summary_Life_Cycle_Costs_and_Carbon_Emissions_of_Wind_Power.pdf

[20]www.parliament.uk/globalassets/documents/post/postpn_383-carbon-footprint-electricity-generation.pdf

[21]data.gov.uk/dataset/9238d05e-b9fe-4745-8380-f8af8dd149d1/solar-photovoltaics-deployment-statistics

[22]www.britishrenewables.com/15mwp-solar-park-will-power-the-uks-first-carbon-negative-business-park/

[23]www.tidalenergy.eu/tidal_energy_uk.html

[24]www.edenproject.com/eden-story/behind-the-scenes/eden-geothermal-energy-project

[25]www.weforum.org/agenda/2017/10/fossil-fuels-will-dominate-energy-in-2040/

[26]www.innovativewealth.com/inflation-monitor/what-products-made-from-petroleum-outside-of-gasoline/

[27]www.mdpi.com/2071-1050/11/8/2276/pdf#:~:text=After%20an%20LCA%20study%2C%20it,CO2e%20per%20km%20of%20road.

[28]www.thelancet.com/journals/lanplh/article/PIIS2542-5196(17)30040-2/fulltext#:~:text=Anaesthetic%20gases%20represent%205%25%20of,acute%20NHS%20buildings%20and%20water.

[29]publishing.rcseng.ac.uk/doi/pdf/10.1308/rcsbull.2020.152#:~:text=The%20carbon%20footprint%20result%20for,CO2e%20per%20hour%20of%20surgery.

[30]www.nationalgeographic.com/science/2018/08/news-helium-mri-superconducting-markets-reserve-technology/

[31]www.steam-museum.org.uk/info/1/steam/9/about_us/2

[32]www.historytoday.com/archive/george-stephensons-first-steam-locomotive

[33]volksrailway.org.uk/history/

[34]www.britannica.com/technology/locomotive-vehicle/Diesel-traction

[35]www.steam-
museum.org.uk/info/1/steam/9/about_us/2#:~:text=The%20completi
on%20of%20the%20last,closed%20and%20sold%20for%20redevelo
pment.
[36]moneyweek.com/403807/11-august-1968-the-last-steam-
passenger-train-in-britain
[37]www.a1steam.com/design/#:~:text=Water%20is%20the%20most%
20significant,mile%20is%20to%20be%20expected.
[38]www.aarp.org/auto/trends-lifestyle/info-2018/how-long-do-cars-
last.html
[39]www.homeserve.com/uk/living/heating-and-cooling/how-long-
do-boilers-
last/#:~:text=The%20average%20life%20expectancy%20for,for%20as
%20long%20as%20possible.
[40] www.twi-global.com/technical-knowledge/faqs/how-long-do-
wind-turbines-last
[41]news.energysage.com/how-long-do-solar-panels-last/
[42]www.guinnessworldrecords.com/world-records/oldest-offshore-
oil-platform-/?fb_comment_id=695118000603625_757999864315438
[43]www.scientificamerican.com/article/nuclear-power-plant-aging-
reactor-replacement-/
[44]en.wikipedia.org/wiki/Ratcliffe-on-Soar_Power_Station
[45]www.bp.com/content/dam/bp/business-
sites/en/global/corporate/pdfs/
[46]www.sciencedirect.com/science/article/abs/pii/S138589471731940
X#:~:text=The%20reverse%20water%20gas%20shift%20(RWGS)%20
reaction%20is%20a%20method,the%20conversion%20of%20CO2.
[47]www.netl.doe.gov/research/coal/energy-
systems/gasification/gasifipedia/ftsynthesis
[48]royalsociety.org/-/media/policy/projects/synthetic-fuels/synthetic-
fuels-briefing.pdf
[49]www.world-nuclear.org/information-library/facts-and-
figures/reactor-database.aspx
[50]www.gov.uk/government/news/pm-outlines-his-ten-point-plan-
for-a-green-industrial-revolution-for-250000-jobs
[51]www.iea.org/fuels-and-technologies/bioenergy

[52]www.fossiltransition.org/pages/post_combustion_capture_/128.php

[53] www.drax.com/technology/how-do-you-store-co2-and-what-happens-to-it-when-you-do/

[54] www.carbonbrief.org/world-can-safely-store-billions-tonnes-co2-underground

[55] www.gov.uk/government/news/pm-outlines-his-ten-point-plan-for-a-green-industrial-revolution-for-250000-jobs

[56]www.electricmountain.co.uk/Dinorwig-Power-Station

[57]www.dominionenergy.com/company/making-energy/renewable-generation/water/bath-county-pumped-storage-station

[58]www.energylivenews.com/2019/12/02/uk-energy-storage-sector-sees-massive-growth/

[59]www.theguardian.com/environment/2019/aug/06/uk-risks-losing-out-europe-home-battery-boom-report-warns

[60] theenergyst.com/the-positives-of-negative-power-pricing/

[61]www.gov.uk/government/groups/heat-in-buildings

[62]www.theccc.org.uk/wp-content/uploads/2018/11/Hydrogen-in-a-low-carbon-economy.pdf

[63]www.enapter.com/hydrogen-clearing-up-the-colours

[64]www.bcg.com/en-gb/publications/2019/real-promise-of-hydrogen.aspx

[65]www.gov.uk/government/news/pm-outlines-his-ten-point-plan-for-a-green-industrial-revolution-for-250000-jobs

[66]techcrunch.com/2019/11/19/heliogens-new-technology-could-unlock-renewable-energy-for-industrial-manufacturing/

[67] heliogen.com/#how

[68] heliogen.com/time-names-heliogen-helioheat-to-list-of-the-best-inventions-of-2020/

[69] theconversation.com/green-ammonia-could-slash-emissions-from-farming-and-power-ships-of-the-future-132152

[70]www.economist.com/media/globalexecutive/black_swan_taleb_e.pdf

[71]www.health.ny.gov/environmental/emergency/chemical_terrorism/ammonia_general.htm

[72]www.royalsociety.org/topics-policy/projects/low-carbon-energy-

programme/green-ammonia/

[73]www.drivingelectric.com/news/678/electric-car-sales-uk-near-7-market-share-september-2020

[74]www.theguardian.com/environment/2019/dec/25/2020-set-to-be-year-of-the-electric-car-say-industry-analysts

[75] www.gov.uk/government/news/pm-outlines-his-ten-point-plan-for-a-green-industrial-revolution-for-250000-jobs

[76]

www.mckinsey.com/industries/automotive-and-assembly/our-insights/charging-ahead-electric-vehicle-infrastructure-demand

[77]wardsintelligence.informa.com/WI058630/World-Vehicle-Population-Rose-46-in-2016

[78]blog.aboutamazon.com/sustainability/go-behind-the-scenes-as-amazon-develops-a-new-electric-vehicle

[79] www.eta.co.uk/2017/09/15/electric-hgvs-with-overhead-power-lines-get-go-ahead/

[80]www.business-live.co.uk/economic-development/final-bill-electrifying-great-western-17948591

[81]www.forbes.com/sites/ericrosen/2018/09/08/over-4-billion-passengers-flew-in-2017-setting-new-travel-record/

[82] www.atag.org/facts-figures.html

[83]www.forbes.com/sites/marisagarcia/2018/10/24/iata-raises-20-year-projections-to-8-2-billion-passengers-warns-against-protectionism/#4b6ae712150f

[84] www.eviation.co/alice/

[85]www.cei.washington.edu/education/science-of-solar/battery-technology/

[86]batteryuniversity.com/learn/archive/comparing_battery_power

[87] simpleflying.com/ryanair-25-minute-turnaround/

[88] www.heathrow.com/company/about-heathrow/company-information/facts-and-figures

[89] www.iea.org/reports/tracking-transport-2019/aviation

[90] www.thetimes.co.uk/article/spaces-for-40-000-cars-under-heathrow-s-green-expansion-78t6m2ghq

[91] www.nbcnews.com/science/science-news/largest-electric-plane-yet-completed-its-first-flight-it-s-n1221401

[92] www.cnbc.com/2020/09/25/hydrogen-powered-passenger-plane-completes-maiden-flight.html

[93] www.airbus.com/newsroom/press-releases/en/2020/09/airbus-reveals-new-zeroemission-concept-aircraft.html

[94] www.statista.com/statistics/564769/airline-industry-number-of-flights/

[95] en.wikipedia.org/wiki/Bell_Boeing_V-22_Osprey

[96] www.sciencefocus.com/space/are-space-launches-bad-for-the-environment/

[97] www.engineeringtoolbox.com/co2-emission-fuels-d_1085.html

[98] www.bbc.co.uk/news/science-environment-49349566

[99] www.space.com/elon-musk-starship-spacex-flights-mars-colony.html

[100] advancedtech.airliquide.com/liquid-oxygen-lox-plant

[101] www.carbonbrief.org/analysis-why-the-uks-co2-emissions-have-fallen-38-since-1990

[102] www.ofgem.gov.uk/ofgem-publications/100628

[103] www.gov.uk/guidance/heat-networks-overview

[104] www.theade.co.uk/case-studies/visionary/islington-councils-bunhill-heat-and-power

[105] www.iea.org/topics/covid-19

[106] www.theguardian.com/world/2020/sep/29/inside-the-airline-industry-meltdown-coronavirus-pandemic

[107] www.thetimes.co.uk/article/coronavirus-biggest-cities-deserted-as-only-17-of-staff-return-to-the-office-0zhslpgz2

[108] www.theverge.com/2020/4/29/21241972/microsoft-teams-75-million-daily-active-users-stats

[109] www.bbc.co.uk/news/business-53925917

[110] www.thetimes.co.uk/article/coronavirus-biggest-cities-deserted-as-only-17-of-staff-return-to-the-office-0zhslpgz2

[111] www.theguardian.com/world/2020/sep/29/inside-the-airline-industry-meltdown-coronavirus-pandemic

[112] www.wired.co.uk/article/frequent-flyer-programme-airlines-environment

[113] assets.publishing.service.gov.uk/government/uploads/system/uploads/attachment_data/file/766109/decarbonising-heating.pdf

[114] www.nationalgrid.com/heating-our-homes

[115] www.theccc.org.uk/wp-content/uploads/2019/07/CCC-2019-Progress-in-reducing-UK-emissions.pdf

[116] energysavingtrust.org.uk/home-energy-efficiency/energy-performance-certificates

[117] www.tado.com/t/en/uk-homes-losing-heat-up-to-three-times-faster-than-european-neighbours/

[118] assets.publishing.service.gov.uk/government/uploads/system/uploads/attachment_data/file/794998/Converting_a_town_to_low_carbon_heating.pdf

[119] www.gov.uk/government/collections/house-building-statistics

[120] carbonintensity.org.uk/

[121] www.nationalgrideso.com/information-about-great-britains-energy-system-and-electricity-system-operator-eso

[122] www.raeng.org.uk/publications/reports/space-weather-full-report

[123] www.ofgem.gov.uk/ofgem-publications/37681/sectoralinvestigations-36.pdf

[124] www.gov.uk/government/publications/updated-energy-and-emissions-projections-2011

[125] www.gov.uk/government/collections/energy-and-emissions-projections

[126] www.itv.com/news/2020-06-28/the-uk-goes-coal-free-for-the-longest-time-in-history-to-create-electricity

[127] www.carbonbrief.org/analysis-uks-co2-emissions-have-fallen-29-per-cent-over-the-past-decade

[128] www.iea.org/reports/renewables-2019/heat

[129] www.gov.uk/government/news/pm-outlines-his-ten-point-plan-for-a-green-industrial-revolution-for-250000-jobs

[130] www.resolutionfoundation.org/comment/after-brexit-the-uk-could-cut-vat-on-energy-but-should-it/

[131] www.tado.com/t/en/uk-homes-losing-heat-up-to-three-times-faster-than-european-neighbours/

[132] heatinginstallerawards.co.uk/2018/04/eu-study-show-energy-saving-potential-thermostatic-radiator-valves/

[133] www.gov.uk/government/news/pm-outlines-his-ten-point-plan-

for-a-green-industrial-revolution-for-250000-jobs
[134]www.tidalenergy.eu/tidal_energy_uk.html
[135]www.gov.uk/guidance/heat-networks-overview
[136] www.thetimes.co.uk/topic/air-pollution
[137]about.bnef.com/blog/battery-pack-prices-fall-as-market-ramps-up-with-market-average-at-156-kwh-in-2019/#:~:text=Shanghai%20and%20London%2C%20December%203,to%20%24156%2FkWh%20in%202019.
[138] www.bbc.com/future/article/20201126-the-solar-discs-that-could-beam-power-from-space
[139] en.wikipedia.org/wiki/Millennials#/
[140]www.ecowatch.com/coffees-invisible-carbon-footprint-1882175408.html#:~:text=Per%20cup%2C%20black%20coffee%20produces,%3B%20each%20latte%2C%20340%20grams.
[141]www.sixdegreesnews.org/archives/10261/the-carbon-footprint-of-a-cheeseburger
[142]publishing.rcseng.ac.uk/doi/pdf/10.1308/rcsbull.2020.152#:~:text=The%20carbon%20footprint%20result%20for,CO2e%20per%20hour%20of%20surgery.
[143] calculator.carbonfootprint.com/
[144] www.eurostar.com/be-en/carbon-footprint
[145]www.carbonbrief.org/analysis-uks-co2-emissions-have-fallen-29-per-cent-over-the-past-decade#:~:text=The%20UK's%20per%2Dcapita%20CO2,or%20the%20US%20(16.6).
[146] data.gov.uk/dataset/9a1e58e5-d1b6-457d-a414-335ca546d52c/provisional-uk-greenhouse-gas-emissions-national-statistics
[147]ourworldindata.org/co2-and-other-greenhouse-gas-emissions